THE COMPLEXITY OF MODERN ASYMMETRIC WARFARE

International and Security Affairs Series
Edwin G. Corr, General Editor

Also by Max G. Manwaring

(ed. with Court Prisk) *El Salvador at War: An Oral History of Conflict from the 1979 Insurrection to the Present* (Washington, D.C., 1988)
(ed.) *Uncomfortable Wars: Toward a New Paradigm of Low Intensity Conflict* (Boulder, 1991)
(ed.) *Gray Area Phenomena: Confronting the New World Disorder* (Boulder, 1993)
(ed. with Wm. J. Olsen) *Managing Contemporary Conflict: Pillars of Success* (Boulder, 1996)
Spain and the Defense of European Security Interests: A Military Capability Analysis (Boulder, 1997)
(ed. with John T. Fishel) *Toward Responsibility in the New World Disorder: Challenges and Lessons of Peace Operations* (Portland, 1998)
(ed. with Anthony James Joes) *Beyond Declaring Victory and Coming Home: The Challenges of Peace and Stability Operations* (Westport, Conn., 2000)
(ed.) *Deterrence in the Twenty-first Century* (Portland, 2001)
(ed.) *Environmental Security and Global Stability: Problems and Responses* (Lanham, Md., 2002)
(ed. with Edwin G. Corr and Robert H. Dorff) *The Search for Security: A U.S. Grand Strategy for the Twenty-first Century* (Westport, Conn., 2003)
(with John T. Fishel) *Uncomfortable Wars Revisited* (Norman, Okla., 2006)
Insurgency, Terrorism, and Crime: Shadows from the Past and Portents for the Future (Norman, Okla., 2008)
Gangs, Pseudo-Militaries, and Other Modern Mercenaries: New Dynamics in Uncomfortable Wars (Norman, Okla., 2010)

THE COMPLEXITY OF MODERN ASYMMETRIC WARFARE

Max G. Manwaring

Foreword by John T. Fishel
Afterword by Edwin G. Corr

University of Oklahoma Press : Norman

Library of Congress Cataloging-in-Publication Data

Manwaring, Max G.
 The complexity of modern asymmetric warfare / Max G. Manwaring ; foreword by John T. Fishel ; afterword by Edwin G. Corr.

      cm. — (International security affairs ;
  p.   v. 8)
 Includes bibliographical references and index.

ISBN 978-0-8061-4265-4 (hardcover) ISBN 978-0-8061-9006-8 (paper) 1. Asymmetric warfare

2. Asymmetric warfare—Case studies.   3. War—Forecasting.   4. Strategy.
5. Insurgency—Case studies.   6. Internal security—Case studies.
7. National security—Case studies.   8. United States—Military policy.
9. Terrorism—Prevention.   10. Military history, Modern.   I. Title.

2011044206

U163.M2687 2012
355.02'18—dc22

The Complexity of Modern Asymmetric Warfare is Volume 8 in the International and Security Affairs Series.

The paper in this book meets the guidelines for permanence and durability of the Committee on Production Guidelines for Book Longevity of the Council on Library Resources, Inc. ∞

Copyright © 2012 by the University of Oklahoma Press, Norman, Publishing Division of the University. Paperback published 2022. Manufactured in the U.S.A.

All rights reserved. No part of this publication may be reproduced, stored in a retrieval system, or transmitted, in any form or by any means, electronic, mechanical, photocopying, recording, or otherwise—except as permitted under Section 107 or 108 of the United States Copyright Act—without the prior written permission of the University of Oklahoma Press. To request permission to reproduce selections from this book, write to Permissions, University of Oklahoma Press, 2800 Venture Drive, Norman OK 73069, or email rights.oupress@ou.edu.

Contents

Foreword, by John T. Fishel	vii
Preface and Acknowledgments	xv
Introduction	3
1. Salient Antecedents to the Present Array of Conflicts: Algeria (1954–1962) and El Salvador (1980–1992)	9
2. New "Kindler and Gentler" Revolutionary Lessons from Peru: The Resurgence of Sendero Luminoso	30
3. Four Trojan Horses of Different Colors: Vignettes from Al Qaeda in Spain, the Cuban Popular Militias, Haiti, and Brazil	51
4. State-Supported Internal and External Persuasion and Coercion: The Russian Youth Group Nashi	76
5. Guatemala at Risk: Drugs, Thugs, and Radical Political Change	97
6. Traumatic Attacks at Another Level: Cyber and Biological War	120
7. The Road Ahead	136

Afterword, by Edwin G. Corr	155
Notes	169
Bibliography	195
Index	203

Foreword

JOHN T. FISHEL

The Complexity of Modern Asymmetric Warfare is the third book in Max Manwaring's invaluable trilogy on conflict in the twenty-first century. While all three books in the trilogy derive from the work Max and his colleagues did on insurgency and counterinsurgency dating from 1984, this set breaks new ground. I am proud to have been associated with Max in his research over that last quarter century and am especially happy that I have the opportunity to make some initial comments on this, his latest work.

The first book in the trilogy, *Insurgency, Terrorism, and Crime: Shadows from the Past and Portents for the Future* (2008), looks at the complex relationship among rebellion, tactics of terror, and organized crime—particularly drug crime—and how those elements intertwine in the real world. In that volume, Max addresses the complicated and complex world of the drug-infested insurgency in Colombia and how it interacts with gangs, drugs, and organized crime in Central America and Mexico. He ties these phenomena in very interesting ways to U.S. immigration policy, noting that the gangs *(maras)* of El Salvador were born in the prisons of California and compounded by the American policy of deporting criminals without informing their countries of origin that these hardened criminal gang members were coming home. By the time the communication issue was resolved, it was too late; the gangs had been transplanted from Los Angeles and San Quentin to San Salvador and were then reimported to Washington, D.C.

The lesson of the gangs is not simply one of crime—organized or not. Rather, it is also a lesson of the consequences of insurgency. The Salvadoran gangs that developed in the California prisons and the streets of Los Angeles were unintended results of migration, legal and illegal, to escape the insurgency that engulfed El Salvador in the 1980s. Because insurgency was also rampant in Guatemala, gang activity developed there for similar reasons and spread to much of the rest of Central America. But as Max suggests, the street gangs of Central America would hardly have become the problem that they did without the combination of insurgency and criminal cocaine cartel activity in Colombia. The current insurgency in Colombia did not begin with the birth of the Fuerzas Armadas Revolucionarias Colombianas (FARC) in 1962 but rather with the start of *la violencia* in 1948. Fighters from *la violencia* were among the founders and leaders of the FARC almost a decade and a half later. The FARC, however, was merely the most effective of several insurgent movements from the 1960s to the 1980s and beyond.

The FARC's Marxist-Leninist organization and Maoist ideology kept it alive through both good and bad decades, all while criminal Colombian entrepreneurs were gaining control of the cocaine smuggling enterprises of South America. In the 1980s, coca was grown almost entirely in Bolivia and Peru and only partially refined in those countries. The partially refined product was bought by Colombian traffickers who finished the refining and ran the distribution to the United States and Europe. Colombian cocaine cartels developed multiple distribution networks to the north—by air, land, and sea. Those networks, particularly the ones on land, required local allies in Central America and Mexico, some of whom were needed simply for their muscle. Enter the gangs of Central America and Mexico.

Then, as Bolivia and, especially, Peru collaborated with the U.S. Drug Enforcement Administration (DEA), Colombian traffickers began to undertake the growing and all the refining of coca into cocaine. As coca growing moved into Colombia, the FARC began to expand into the same areas of the country. In response, cocaine traffickers and cattlemen (there often was a high degree of overlap) began to support armed militias to defend their properties from the FARC. The militias, called by many paramilitaries, were both tacitly and actively supported by the Colombian Army and began to counterattack the FARC (and other Marxist guerrillas as well). Major government efforts, with U.S. assistance, to dismantle the cocaine cartels had enough success to produce at least two important unintended consequences: (1) Mexican cartels began to take over the cocaine distribution system, and (2) the FARC went into the cocaine production business—sometimes in partnership with some paramilitaries, sometimes with smaller organized criminal elements, sometimes in both political and economic competition with the same. All of this Max ties together in *Insurgency, Terrorism, and Crime.*

Another aspect of the book is his focus on historical situations. In rough order from start date, Max addresses the Tupamaro insurgency in Uruguay that began in 1962, was defeated, and then came to power as a responsible political party after 2000; the Italian Red Brigades terrorist movement that began in 1968 and was defeated by 1983; and the Portuguese military revolution of 1974 that overthrew the dictatorship and put Portugal on the road to a democratic future. The common element of these three cases is the long-term success of democratic reform coupled with appropriate tactical efforts to defeat antidemocratic revolutionaries and set the countries on a path toward political freedom with stability. In consonance with this conclusion, the cases demonstrate one possible future for Colombia, Mexico, and Central America. While alternative, and less desirable, futures are suggested by the chapters previously discussed, the cases of Venezuela and Al Qaeda point toward additional challenges of asymmetric warfare facing the United States in the next decades of the twenty-first century.

The second book of the trilogy, *Gangs, Pseudo-Militaries, and Other Modern Mercenaries: New Dynamics in Uncomfortable Wars* (2010), picks up where the previous book leaves off. Using Lenin as a foil to introduce a strategy of revolution, Max examines the FARC and the cartels of Colombia and Mexico as practitioners of Leninist methods. Moving out from these themes (explored in the previous volume from different angles), he addresses the use of "gangs" by state actors to both expand and maintain control of the people. Venezuela again appears, but Hugo Chavez is portrayed more clearly as an individual who is slowly expanding his dictatorship through the use of these bought-and-paid-for mobs. This theme is also explored with respect to Argentina.

Max's last case in this volume is that of Al Qaeda's "franchise" gangs in Europe.

I comment on Max's achievement in my afterword for this second book, where I stress that he has rediscovered a phenomenon that dates from ancient times in the West. Indeed, all of these asymmetric threats have been seen in analogous form before—in Roman times, in British India, and in the Union occupation of the South after the American Civil War, to name a few. Analogies, however, are not the same things happening over and over again. Rather, they are somewhat similar things that appear in different times and places, posing problems that are only somewhat similar to those seen before. One of the more difficult problems facing modern statesmen, soldiers, and cops is figuring out where an analogy applies and where it ends. Thus, whereas the second work calls attention to what is analogous, it is in the third that we find how these threats have transformed into something that is both old and new at the same time.

Max begins this volume with a step into the past. He addresses two insurgencies that had substantially different outcomes: Algeria and El Salvador.

In the first case, Algeria, the French very successfully employed an enemy-centric strategy to the point at which they had, to all intents and purposes, destroyed the insurgents inside the boundaries of Algeria. The French general that Max and his colleagues interviewed was adamant that they had not been defeated. Yet when one of the interviewers asked the general whether it was still "Algerie Française," the answer was obviously not, and so the general was forced to admit that France had lost the war. The situation is reminiscent of U.S. Colonel Harry Summers's encounter with a North Vietnamese colonel during the negotiations about repatriation of American prisoners. Summers said that the North Vietnamese had never beaten the Americans on the battlefield. The Vietnamese colonel allowed as how that may have been true but it was also irrelevant—the Vietnamese had won. Algeria (and Vietnam) demonstrates that in a major insurgency, one cannot kill one's way to victory; however, as Sir Robert Thompson has pointed out, one cannot defeat an established insurgency without killing.

The El Salvador case clearly illustrates both points. The government of El Salvador and the El Salvador Armed Forces (ESAF) very nearly lost their war by trying to kill their way to victory. Only when reform and democracy took effect and intelligence provided real insurgent targets who could be attacked, captured, or killed did the Salvadorans defeat the insurgency. However, the most interesting aspect of that victory is that the Salvadoran government never claimed it; they negotiated an end to the conflict on the terms President Duarte offered in 1985 but agreed with the insurgents that peace had come, and with it, a new era. The proof of a successful peace has been the recent election to the presidency, without bloodshed or major social disruption, of the political party that grew out of the insurgency. In this outcome, we see a situation analogous to Uruguay, which Manwaring addresses in the first book of the trilogy.

Max follows his discussion of past insurgencies with a focus on a present one arising, like a phoenix, out of the ashes of the past. In this case, he considers the resurgence—such as it is—of Peru's Sendero Luminoso (SL). The original SL largely dissolved when its leader, Abimael Guzmán, was captured in 1992. However, two elements survived. One was the Huallaga Regional Command in the coca-growing country of the upper Huallaga Valley, while the other was made up of elements that refused to disband and continued to operate in the SL heartland of Ayacucho. As the Huallaga Regional Command followed the path of Colombia's FARC and validated the self-fulfilling prophecy that SL was financed by the drug trade (untrue before 1992),[1] the Ayacucho-based elements reconstituted and took advantage of the fact that nothing had really changed for the indigenous people of the southern Andes in and around Ayacucho. Moreover, these elements adapted tactically, discarding some of their more brutal past behavior. The result has been a resurgence of an insurgent threat from SL, now more clearly funded by the illicit drug trade. This, Max identifies as a blurring of the lines between insurgency and criminality, precisely what was found in Colombia and reported in both of the prior volumes of the trilogy.

At this point in the story, Max shifts direction and begins to look at several different manifestations of the "uncomfortable wars" phenomena. These he calls "Trojan horses" of different colors, and he finds commonalities among Al Qaeda in Spain, the popular militias that sustain the Cuban government, thuggish militia-like elements in Haiti, and Brazilian drug gangs in the *favelas*. What is common to all of these disparate organizations is their need to control territory to the degree necessary to sustain freedom of action. Although none of them seek to overturn the government and assume the responsibility for governing, they all need the government to look the other way while they are conducting their business. In striving for this freedom of action, they all challenge the sovereignty of the de jure government and its monopoly of armed violence. All are forms of "counter-states."

From this initial analysis of the Trojan horses, Max delves deeper into the phenomenon with his focus on the Russian youth militia (the Nashi) and the gangs of Guatemala. The Nashi are similar to the Cuban militias in that they are a partially controlled arm of the state. Because state control is only partial, however, the organization represents as much a threat to state security as it does a support. This paradox remains unresolved.

With the criminal gangs of Guatemala, Max returns to a major theme of the first two books of the trilogy. Organized criminality has much in common with insurgency, yet it is not the same. The Guatemalan gangs, like their brothers in Mexico, El Salvador, and Colombia, do not seek the overthrow of the state. In fact, they benefit too much from the state-provided public services; having to provide such services would cut much too deeply into the gangs' profits. It is, therefore, far cheaper to rent the politicians for those actions that will improve the gangs' profits and let the government do the heavy lifting on the "public goods." The problem with this approach, of course, is that government this corrupt soon loses its capabilities to provide needed and desired public goods—especially security—and thereby forfeits its legitimacy.

One potential outcome of both the state-supported militias and the criminal gangs, especially in our globalized world, is migration to places of greater security and opportunity, such as the United States. We have seen, in the past year or so, a major intelligence effort by the Russian Security Service, the SRV, to penetrate American economic and political security. Was the Nashi either involved directly or a source of Russian agents? Similarly, with the Guatemalan gangs—and their Latin American brothers—there is much penetration of the United States through extensive illegal immigration, drug importing, and the smuggling of people. That, plus the fact of the American street and prison origins of MS-13 and the Mara Calle 18, as well as other similar gangs from around the region, points to a significant threat to American security, especially when we recall that the objective is not governmental overthrow but rather sufficient control to guarantee freedom of action and movement.

As potentially dangerous as all of this is, consider how much more dangerous it all could be if it were coupled with such additional threats as cyber and/or biological warfare. There is nothing new about biological warfare: Thucydides recorded the effects of the plague on Athens under siege during the first years of the Peloponnesian War. Nobody knows whether the plague was induced or simply happened, but it was clearly an effective ally of the Spartans, robbing Athens of many of her best leaders, including the incomparable Pericles, as well as significant numbers of her fighting men. In 1763, during the Pontiac Conspiracy, the British attempted to infect their Indian adversaries by distributing blankets that had been used by men and women ill with smallpox. Then, in 2001, shortly after the September 11 attacks, letters filled with anthrax powder began to turn up from Florida to Washington, D.C. Although only five people were killed, the anthrax attacks caused millions of dollars in lost time and production as well as in efforts devoted to finding the perpetrator.

The newest threat to appear over the horizon is that of cyber warfare. Clearly, cyber-attacks can be instituted by governments either overtly, covertly, or clandestinely—all of which have been used in the immediate past. There have been cyber-attacks that originated in Russia against both Estonia and Georgia, with the latter coinciding with the 2008 "war" between the two over South Ossetia and Abkhazia. There have also been numerous cyber-attacks against U.S. Department of Defense systems, as well as at least one against Google that originated in China. Whether any of these were government policy is not totally clear, but given the nature of the states involved, that is hardly an unreasonable inference. There is also the case of the Stuxnet attack on Iran's nuclear program, which is usually attributed to Western intelligence agencies, especially the United States and Israel.

Perhaps even more threatening is the use of cyber warfare by non-state actors such as Wikileaks, which orchestrated attacks on Master

Card and Visa for cutting payment services to their organization. What was particularly interesting was the decentralized nature of the attackers, which appears to be analogous to the now more decentralized "new and improved" Al Qaeda in its terror operations.

This review of Manwaring's trilogy points to his major achievement. In the tradition of the intelligence analyst that he has been for much of his career, Max has examined all manner of threats that face the world of the twenty-first century. Although he has not emphasized traditional state-to-state conflict, it is addressed in part with respect to cyber war either alone or as an additional dimension to conventional encounters.

The main focus of the trilogy, however, is on the wide variety of asymmetric threats extant. As Max is fond of pointing out, not much of this is new; what is new is the way in which these threats manifest themselves. In this, he has responded to his favorite admonition from Carl von Clausewitz: "'The first, the supreme, the most far reaching act of judgment that the statesman and commander have to make is to establish . . . the kind of war on which they are embarking; neither mistaking it for, nor trying to turn it into something that is alien to its nature.' Determining the nature of the conflict is thus 'the first of all strategic questions and the most comprehensive.'"[2] Thus, Max's achievement is to have determined the nature of conflict in this still-new century. I commend to you this third volume of his trilogy.

Preface and Acknowledgments

This book explores an important and meaningful aspect of contemporary and future irregular asymmetric revolutionary conflicts. Because it takes probable enemy perspectives into consideration, it serves as a beginning point for decision makers, policy makers, planners, and anyone else who might have the responsibility for dealing with, analyzing, or reporting on contemporary hegemonic nonstate actors who do not rely on highly structured organizations, large numbers of military forces, or costly weaponry.

In any event and in any phase of a hegemonic process, violent nonstate actors have played and continue to play substantial roles in helping their organizations or political patrons coerce radical political change and achieve putative power. These goals and actions define insurgency as well as war and shift the irregular asymmetric global security challenge from abstract to real. Accordingly, this book examines cases that illustrate how would-be hegemonic political actors all around the world might seek to realize their utopian dreams.

This book is the direct result of a U.S. Army War College Faculty Research Grant that allowed me the time to explore lessons learned from cases that might be useful in illustrating models of harbinger types of hegemonic nonstate actors' efforts to achieve political, commercial, and/or ideological objectives. For that, I am happy to thank Professor Douglas C. Lovelace, Jr., director of the Strategic Studies Institute at the

U.S. Army War College, and Major General Greg F. Martin, commandant of the U.S. Army War College.

I would also like to thank the "usual suspects" whose knowledge, expertise, good sense, and patience helped make this book possible. For that, I must begin with Ambassador Edwin G. Corr and Lieutenant Colonel John T. Fishel, U.S. Army, retired. They have steadfastly pushed, pulled, and otherwise encouraged me in this initiative. A few others should also be acknowledged. They include Colonel Robert M. Herrick, U.S. Army, retired; Ambassador David Passage, retired; and my colleagues Robert Bunker, Gabriel Marcella, and William J. Olson. They all have—in their own ways—been most kind and helpful. Then, there is the gifted Ms. Sally Bennett. She has done an absolutely outstanding job of copyediting this trilogy and has made me look a lot more articulate and intelligent than I really am.

In this context, I respectfully and lovingly dedicate this book to my wife, Janet. She has taken turns with Ed Corr and John Fishel at various times, on different occasions, and in different ways to beat me about the proverbial head and shoulders and prod me onward.

This book should not be construed as reflecting the official positions of the U.S. Army War College, the Department of the Army, the Department of Defense, the Department of State, or the U.S. government. I, alone, am responsible for any errors of fact or judgment.

The Complexity of Modern Asymmetric Warfare

Introduction

Globally, there are well over one hundred ongoing small, irregular, asymmetric, and revolutionary wars,[1] in which violent nonstate actors are helping their own organizations or political patrons bring about radical change or acquire power. Because of the ubiquity of this asymmetric global security challenge, strategic leaders will be grappling with an increasing number and increasing variety of security problems in the twenty-first century, and they must be prepared to think about these issues from multiple angles, at multiple levels, and with varying degrees of complexity.

Thus, the research objectives of this book are to (1) familiarize strategic leaders with the complexity and effectiveness of contemporary irregular conflict and gain new insights into it; (2) articulate the proven directions and paths nonstate actors have taken and are taking toward achieving national, regional, and global hegemonic objectives, as well as illustrate effective and ineffective practices in dealing with them; and generally, (3) help civilian and military leaders to think strategically about the many "wars amongst the people"[2] that have emerged out of the cold war and are significant forerunners to many of the irregular wars that the United States and its allies face now and into the future.

In these terms, it is important to take probable enemy methods of attack into consideration. The end product is a book elaborating harbingers that are considered by practitioners and theorists to be most relevant now and for the future.[3] Accordingly, this book examines cases that illustrate how would-be revolutionaries all around the world, such as the following, might seek to realize their dreams:

Populists and neopopulists
The New Left, New Socialists, or twenty-first-century socialists
Criminal nonstate actors
Agitators, gangs, popular militias, and other organizations used by state and nonstate actors

THE CHANGING FACE OF WAR

Since Venezuela's President Hugo Chavez distributed Jorge Verstrynge's *Peripheral War* to his and several other officers from the international community, it has become one of the most comprehensive and universally read analyses of contemporary revolutionary conflict. Chavez admonished everyone to learn from this work and develop doctrine for fourth-generation (asymmetric) war. Verstrynge, a Spanish Leninist scholar, calls contemporary conflict *guerra periferica* (peripheral war). It is a combination of asymmetrical, unrestricted, and total war, and what Sun Tzu and Mao Tse Tung might have called indirect and protracted war.[4] Its defining characteristics are

A "revalidation of guerrilla war," that is, making guerrilla war less military-focused and more multidimensional, to include the use of time as an instrument of power;
The "deterritorialization and denationalization of conflict," that is, one's focus on physical territory shifts to the human territory and use of neighboring and other regional countries for support, resources, and sanctuaries;
The use of communication and information, with the media becoming a primary instrument of power;
The "relativization of the time factor" in contemporary conflict, that is, less concern regarding the achievement of a quick and easy military victory and more with the long war (prolonged/protracted war);

The placement of relatively small groups of combatants, agitators, or civilians interspersed among ordinary citizens, with no permanent locations and no identity to differentiate them clearly from the rest of a given civil population, concealing the threat they pose (possible "Trojan horses") and enabling them to create and use devastating surprise ("divine surprise") as a major instrument of power; and

The eruption of Islam as a valid and competing ideology in the global security arena.[5]

As an example of the level of acceptance of these characteristics of peripheral war, these are the principal characteristics of what president Hugo Chavez of Venezuela now calls "asymmetric war," "fourth-generation war," "people's war," or "super insurgency."[6] These characteristics of contemporary conflict provide hard evidence that war is changing. That requires a change in the conceptualization of warfare, which, in turn, demands fundamental change in how conflict is managed. That will require the astute political-psychological use of "brain power" rather than the traditional application of technology and "brute force."

METHODOLOGY AND CASES

The case study method is ideal for the purposes of this book. Researchers who utilize the methodology frequently find that the study of only a few sharply contrasting instances can produce a wealth of new insights. Additionally, this research approach leads to an enhanced understanding of the architecture of successful or unsuccessful strategies—or best/worst practices—in dealing with small, complex, irregular wars. Thus, the purposive sample of cases outlined below was chosen for the three specific research purposes listed above. In this context, the environment within which the conflict takes place and the various activities undertaken to conduct or counter it serve as independent variables. The end result acts as the dependent variable. With this information, strategic analytical commonalities and recommendations can be determined for strategic leaders that are relevant to each specific type of conflict explored, as well as the larger general phenomenon.

This takes us to the theoretical linear-analytic approach to this book. Robert K. Yin defines the linear-analytic approach as the traditional or standard approach to case studies. That is, the major components include "the issue, the context, the findings, and conclusions and implications." The issue and context answer the "what and why" questions; the findings examine the "who, how, and so-what" questions; and conclusions and implications address key points, recommendations, and countermeasure issues.[7] Consequently, the primary components of each case study are the following: issue, context (theoretical, historical, political, economic, sociopolitical, organizational), findings or outcome (where the protagonist's program leads or has led), and key points and lessons.

These elements are closely related and overlapping, in that they are mutually influencing and constitute the "cause and effect" dynamics of a given situation. Without a fundamental understanding of the answers to these questions, the various types of state-supported, state-associated, or independent gang activity are not likely to be clearly perceived, and the response to such activity may turn it into something that it actually is not.[8] The cause and effects related to the components of the linear-analytic approach, as discussed in the following chapters, demonstrate that associated threats are not abstract—they are real. The primary intent of this methodology, then, is to help political, military, and opinion leaders, as well as concerned citizens, think strategically about explanations of many of the "wars amongst the people" that have emerged out of the cold war and are taking us kicking and screaming into the twenty-first century.

More specifically, this book examines a few premier complex, irregular, asymmetric, peripheral war cases that illustrate the following: antecedents to the present array of conflicts, lessons from a "kindler and gentler" guerrilla war, vignettes of "Trojan horses," state-supported persuasion and coercion, drugs and radical political change, and cyber and biological warfare. The most salient reality demonstrated in these cases is that many state-supported, state-associated, and independent nonstate actors all around the world are deeply involved in indirect (peripheral) struggles to coerce radical political change—and achieve or maintain political power. These actors are being ignored or, alternatively, considered too hard to deal with. Yet they have the capability to seriously threaten the interests and well-being of the global community.[9]

The contemporary internal security situation is, thus, further characterized by unconventional battlefields that no one from the traditional-legal Westphalian school of conflict would recognize or be comfortable with. In addition to conventional *inter*state war conducted by uniformed military forces of another country, we see something considerably more complex and ambiguous.

CONTEMPORARY BATTLEFIELDS

Thanks to Steven Metz and Raymond Millen and their theory-building efforts, we see a new and broadened typology of contemporary battlefields that includes (1) conventional direct state versus state conflict; (2) direct and indirect nonstate actor versus state conflict; (3) indirect state versus state conflict using surrogates or proxies; and (4) *intra*state conflict involving nonstate actors versus other nonstate actors and the state.[10] Regardless of any given politically correct term for war or conflict, all state and nonstate actors involved in any one or more of the four contemporary battlefields noted above are engaged in one common political act—war. That is, the goal is to control or radically change a government and to institutionalize the acceptance of the victor's will.[11]

In this fragmented, complex, and ambiguous political-psychological violence-dominated security environment, conflict must be considered and implemented as a whole. The power to deal with these kinds of situations is no longer hard combat firepower or even a more benign police power. Rather, power consists of the multilevel, combined political-psychological, moral, informational, economic, social, police, and military activity that can be brought to bear holistically on the causes and consequences—as well as the perpetrators—of violence.[12]

The logic of that situation argues that the conscious choices that strategic civil-military leadership in individual nation-states and the international community make about how to deal with the contemporary nontraditional threat situation will define the processes of national, regional, and global security and well-being now and far into the future. Thus, the strategic purpose of this book goes beyond an attempt to address a major gap in the international relations and hegemonic war literature. The fundamental relevance and imperative of this book lie in the transmission of hard-learned lessons of the past and present to current and future leaders. Each case has something to teach. There is no single defining way of war: it changes over time and with experience. In every case, however, taking a somewhat different enemy into consideration is of critical importance, and the practical imperative of this book is to give the reader tools to do just that.

CHAPTER 1

SALIENT ANTECEDENTS TO THE PRESENT ARRAY OF CONFLICTS

Algeria (1954–1962) and El Salvador (1980–1992)

Once again, as after the ending of the cold war, political and military leaders all around the world are in the process of rethinking the global role and strategies of their countries in terms of military reform and the restructuring (transformation) of their armed forces to better deal with the security problems that are likely to arise over the next several years. Strategic consideration of the possible enemy has so far played little part in the debate, which has focused generally on the tactics and operational minutiae of contemporary conflict.

Before asking the usual questions—"What are we going to do?" "How is it to be done?" "Who is going to command and control the effort?"—the statesman and the commander (as Clausewitz termed them) must first "establish...the kind of war on which they [are] embarking."[1] Determining the nature of the conflict is not only the first but also the most comprehensive of all strategic questions.[2] The central strategic problem, then, in complex irregular conflicts is to determine the primary center of gravity and prioritize the others. Clausewitz teaches us that in these kinds of conflict the primary center of gravity is no longer the enemy military force or the enemy's capability to make conventional maneuver war. Rather, the primary center of gravity has shifted to public opinion and leadership.[3] Strategic civilian and military leaders and their staffs face the challenge of understanding what the enemy is attempting and how he intends to do it; as a result, in contemporary conflict, strategic leaders must correctly identify the primary center of gravity, prioritize the others, and link policy, strategy, force structure and equipment, and campaign plans to the kind of war on which they are embarking. A sure formula for generating a national disaster is fighting a war one is prepared for and comfortable with against an enemy that is operating with completely different perceptions and strategic objectives, such as fighting a conventional war of attrition against an enemy who is more subtly preparing to take control of a targeted population.

Thus, the intent of this chapter is to help political, military, and opinion leaders understand the differences and implications of the many conflicts that have emerged out of the cold war and are taking us kicking and screaming into the twenty-first century. We explore two cases that can be considered salient antecedents to the present array of conflicts taking place around the globe—those of Algeria (1954–62) and El Salvador (1970–92).

CASE STUDIES

The Algerian War for Independence

The Algerian case illustrates the beginning of the end of purely military-oriented conflicts and the beginning of the use of a multidimensional approach (combinations) to revolutionary war. This is not a new approach to conflict, however. Vladimir Ilyich Lenin advocated the use of *all* instruments of power in his revolutionary rhetoric. Moreover, he reserved the use of armed force for the ending of a conflict, after the bourgeois enemy was demoralized, exhausted, and ready to capitulate.[4] Thus, Lenin applied the notion of "total war" to "defensive war" on two different levels: first, one must achieve total political objectives (radical change); and second, one must use a totality of ways and means to compel fundamental political change and new (socialist) values.[5]

This may be accomplished by individuals with minds that are flexible and familiar with the indirect approach (for example, Sun Tzu, Lenin, Liddell-Hart, and Verstrynge). These individuals must also be familiar with the power of dreams and perceptions and must have a grasp of the history of the unconventional and of the various possible mixes of military and nonmilitary, lethal and nonlethal, state and nonstate political-psychological actions.[6] The effectiveness of these talented individuals, however, is dependent on end-state planning and an authoritative and integrated strategic implementing organization and process. Moreover, the Algerian case shows us that approaches to war can change.[7] War has become a juxtaposition of small, disparate but highly interrelated actions (dimensions). The combined power of diplomacy, intelligence, ruse, media (propaganda) manipulation, and physical coercion has succeeded the limited power of the "shock and awe" of conventional maneuver warfare. Thus, the Algerian case confirms the notion that the concept of "enemy" and the conventional military center of gravity are drastically changed and the ways and means of attacking an opponent infinitely broadened.

The Salvadoran Insurgency

The case of El Salvador represents a Vietnamese-Chinese model of guerrilla war made popular in Latin America by Ernesto (Che) Guevara. Che Guevara took exception to the wisdom of Lenin and the experience of the Algerians regarding the ways and means necessary to conduct a revolution. He argued that an insurrection, in itself, would create the conditions necessary to generate a successful revolution. Such an insurgency could be organized by a relatively small but mobile group of guerrilla fighters—that is, the *foco*.[8] A targeted government's inability or unwillingness to eliminate or nullify the insurgents would force it to overreact against the rebels and in so doing would begin to alienate the people. The foco would, then, act as a catalyst to challenge the legitimacy of the incumbent government and initiate popular support for the "revolution."[9] This thinking was and remains deeply engrained in general revolutionary thinking. As one example, early in the insurgency, some high-ranking FMLN (Farabundo Martí National Liberation Front) leaders argued that organized violence is not only a shortcut to radical political change but also the only way to "achieve the power" to make the profound changes needed in the targeted society.[10]

Later on in the conflict, FMLN strategy was altered from time to time in recognition of changing political-military conditions, the earlier Vietnamese experience in Southeast Asia, and the Algerian Revolution. In the end, after a negotiated settlement in 1992, the FMLN insurgents were incorporated into the democratic Salvadoran political process.

THE WAR FOR ALGERIAN INDEPENDENCE, 1954–1962

In Algeria, the stage was set for an attempt at revolution early in the twentieth century. It was organized by a puritanical Islamic sect that called for a total rejection of European culture and physical isolation from the French. Subsequently, the revolution of the 1950s and early 1960s was not something that came out of the desires and needs of the peasantry and proletariat. And it was not something that represented the Algerian people's deepest ideological concerns. Whatever resistance to French rule as there had been from about 1830 to 1940 was conducted almost exclusively in the name of religion, not nationalism. Moreover, that resistance, after World War II, was orchestrated and led by a small group of middle-class intellectuals and politicians. A notable example of the elite nature of the Algerian revolutionary leadership was Ferhat Abbas. Educated in Europe and a great admirer of French culture and power, Abbas believed that before the arrival of the French, no such thing as an Algerian nation existed. When the French arrived in the area, there was no nation, no national history, no boundaries, no government, and no name.[11]

Historian Matthew Connelly argues that the origins of the grand strategy for the Algerian War can be traced to the last day of World War II. Nationalists all over the world, including Algerians, had associated themselves with American anticolonialism and organized celebratory marches. Many of these, including the Algerian marches, quickly turned into clashes with colonial forces, in which French forces reportedly killed from 6,000 to 45,000 Algerians. In turn, Algeria's leading political opposition figure, Messali Hadj, created a political party (the Mouvement pour la Triumphe des Libertes Democratiques, Movement for the Triumph of Democratic Liberties [MTLD]). The party won municipal elections all across Algeria, but in the 1948 elections the French arrested MTLD candidates and stuffed ballots for their "Muslim yes-men."[12]

Later that year, the MTLD asked the head of its paramilitary section, Hocine Ait Ahmed, to advise the party on how it might win Algeria's independence through the use of armed force. He studied carefully previous examples and the insights of Clausewitz and Liddell-Hart and came to the conclusion that the use of force could not succeed against the French in Algeria. Alternatively, Ait Ahmed prescribed a strategy that would coordinate a combination of guerrilla war, information war (propaganda and public opinion), and diplomatic war. The primary effort, however, would be a diplomatic (foreign policy) war of national liberation. The MTLD approved the report in December 1948. Younger militants, who were not particularly happy with a strategy that reduced armed force to a supporting role, created the Front de Liberation Nationale (National Liberation Front [FLN]). In 1954, that organization accorded the military dimension the same status as the diplomatic component of the war of national liberation. In the end, by 1962, the French had completely destroyed the Algerian military effort. Algerian independence was achieved by a strategy that aimed at establishing a mutually reinforcing relationship between the diplomatic campaigns abroad, and efforts to generate popular support at home and abroad.[13]

The following commentary is a very brief exploration of the overwhelming need to think and act globally to attain one's objectives at home and abroad. Moreover, it is the position of myself and analysts with whom I have worked closely, along with that of Qiao Liang and Wang Xiangsui, that the use of combinations of instruments of power is key to success in conducting contemporary insurgencies and counterinsurgencies.[14] This requires the same strategic considerations as outlined above for the El Salvadoran conflict: war is total and requires a multidimensional whole-of-government approach; the fundamental objective of war is political, not military; and to deal with the entire phenomenon, special organizations, strategies, and ideologies are necessary. In that connection, the French in Algeria had it right. To win an insurgency or counterinsurgency war, one needs a viable ideology and strategy, as well as an organization strong enough to generate a unity of effort. The problem was that the French did not have the discipline and organization to develop those requirements. Roger Trinquier clearly points out that the Algerians did not develop those requirements.[15]

Political Context

France's Fourth Republic, which lasted until General Charles De Gaulle took power in 1958 and created the Fifth Republic, was handicapped by its parliamentary and electoral systems. These systems favored multiple political parties and produced unstable coalition governments. During the whole period of its tenure after the end of World War II, the Fourth Republic averaged a new government every eight months. Moreover, just before the promulgation of the Fifth Republic, France's third government in eleven months resigned on April 15, 1958. The political instability and governmental inertia in the period after the end of World War II caused, as only a few examples, the loss of two-thirds of France's currency reserves, a growing foreign debt, increasing budget deficits, and chronic inflation. Clearly, the French economy and the French people had been ill served by its political system. Likewise, French troops participating in the colonial wars after the end of World War II had been ill served by their political masters in the Fourth Republic and were beginning to exhibit mutinous attitudes toward their political leaders. Additionally, Algeria had a large European population that had considerable political clout in the French parliament and was the cause of governmental deadlock on more than one occasion. Thus, the political, economic, military, and social disarray in postwar France generated a situation in which neither the government nor the army could develop a viable ideology, strategy, or the discipline and organization to implement the necessary components of an effective counterinsurgency.[16]

All was not perfect harmony in North Africa, either. The Algerian insurgent organization was split by jealousies and rivalries both within Algeria and between supposed external allies. Inside Algeria there were both proponents of a strong military war to win independence and advocates for a political-informational-diplomatic war against the French.[17] Externally, as an example, "[t]he Tunisians joined the Moroccans in urging the [Algerian revolutionaries] to accept less than full independence."[18] Nevertheless, all the parties to the revolutionary process appear to have agreed that the Algerian conflict was a kind of world war for public opinion. This, then, was the unifying element that—despite military defeat at the hands of the French—kept the Algerian revolutionaries more or less together and led to the Evian Accords of March 1962 and independence on July 4, 1962.[19]

Ait Ahmed's Strategy

Diplomatically, the Algerians played superpower against superpower. They exploited every international rivalry that offered potential leverage, including revisionist against conservative Arab states, the Arab League against Asian neutrals, China against the Soviet Union, the communist powers of Eastern Europe against the Western NATO allies, and the United States against France. The important point, however, is that no amount of diplomatic virtuosity would have sufficed if the Algerian insurgents' activities abroad had not resonated with the people they represented—and global public opinion. The genius of Ait Ahmed's program for diplomatic war, then, was to ensure that the diplomatic, public opinion (information), and military components of the campaign had high-level (strategic) political objectives, were mutually reinforcing, and were communicated effectively around the world.[20]

The Algerians also exploited French mistakes in Algeria. By the end of 1957, the French had almost completely eliminated terrorism in the large towns, effectively interrupting transit between Algeria and its external support. Nevertheless, some of the methods used in achieving those objectives (such as torture, censorship, and attrition) were reprehensible, and the French Army managed to alienate much of the Algerian population, French public opinion at home, and world opinion and leadership. Algerian international and public diplomacy and information campaigns at home and abroad were provided material they could not have invented. It was only a matter of time before virtually everyone in the global community clearly understood that French moral rectitude was deficient and the Algerian alternative was far superior.[21]

As a consequence, Algiers and Washington, New York (the United Nations), Beijing, Moscow, Paris, and other capitals around the world became theaters in the same war. What happened in these places, many thousands of miles from Algeria, turned out to be more decisive than what was happening on the conventional Algerian battlefield. Jean Larteguy caught the essence of French frustration regarding contemporary conflict in *The Centurians:*

> It is difficult to explain exactly, but it is rather like (the card games) Bridge as compared to *Belote*. When we (the French) make war, we play *Belote* with 32 cards in the pack. But the [insurgent's] game is bridge and they have 52 cards: 20 cards more than we do. Those 20 cards short will always prevent us from getting the better of them. They've got nothing to do with traditional warfare, they're marked with the sign of politics, propaganda, legitimacy, agrarian reform.... What is biting [the French officer]? I think he is beginning to realize that we've got to play with 52 cards and he does not like it at all. Those 20 extra cards aren't at all to his liking."[22]

Human Terrain

The uncomfortable reality of contemporary asymmetric, irregular, unconventional war is that it is now a vast interlocking system of actions—political/economic, social/moral, psychological/informational, and military/paramilitary—that aims to overthrow the established authority in a country or other geographical region, or some specific human terrain. Nevertheless, even though there is a certain concern with the physical terrain, the human terrain is the primary key to the struggle. In view of the present-day interdependence of the global community, any residual grievance within a targeted terrain, no matter how localized and lacking in scope, will surely be brought by determined adversaries into the framework of the great world conflict of public opinion. General Sir Rupert Smith argues succinctly that one of the major characteristics of modern conflict is that the major military and nonmilitary battles take place among the people; when they are reported, they become media events that may or may not reflect social reality.[23] But that is really not new; Clausewitz warned us almost two hundred years ago that in this kind of war, the primary center of gravity is no longer an enemy military formation or the industrial-logistical ability to conduct maneuver warfare. It is public opinion and leadership.[24]

Similarly, Qiao and Wang teach us that although there may be a primary or dominant center of gravity (dimension/domain)—regardless of whether a war took place 2,500 years ago or last year—the evidence indicates that all victories display one common denominator. That is, the winner is the power or power block that best combines the primary elements of statecraft. Within the context of combinations or collective activity or multidimensional activities, understanding that there is a difference between the "dominant" sphere and the "whole" is crucially important. There is a dynamic relationship between a dominant type of war (for example, military, economic, or moral) and the supporting elements that make up the whole. As an example, military war should always be supported by media (information) war and a combination of other types of war such as economic war, cyber-network war, and diplomatic war. At base, however, the intent of every type of war—with its dynamic combinations of multidimensional efforts—is to directly support the main strategic political objective of the entire effort.[25]

The combining of multiple dimensions and subdimensions provides considerably greater strength or power than one or two operating by themselves. This gives new meaning to the idea of a protagonist using *all* available instruments of power at his disposal. This, then, requires the understanding of warfare as a whole, the need to develop an ideology (theory of engagement), and the need to develop organization and doctrine to enable a comprehensive unity of effort.[26]

Key Points and Lessons

The Algerian war for independence is instructive in many ways. The most important, however, would include the following.

- Modern conflict is not a test of expertise in creating instability, killing enemies, conducting reprehensible violence, or achieving commercial, ideological, or moral satisfaction. Instead, it is

an exercise in survival. Failure in contemporary unrestricted combination warfare is not an option.
- Algeria is a classic case study in the superiority of nonkinetic versus kinetic approaches to contemporary war. Algeria is also the first case over the past several years to clearly demonstrate that political-military leaders at all levels must understand how force can be employed to achieve political-psychological ends and the ways in which political-psychological considerations affect the use of force.
- An insurgency or counterinsurgency (two sides of the proverbial coin) cannot be conducted in an ad hoc, piecemeal, or "business-as-usual" (lackadaisical) manner. A viable ideology (theory of engagement), the adroit use of combinations of available means, and an organization at the highest level that can ensure a complete unity of effort are absolutely necessary.
- The Algerian case explicitly and implicitly demonstrates that some of the most important elements of contemporary statecraft include public diplomacy at home and abroad, intelligence, information and propaganda operations, cultural manipulation measures to influence or control public opinion and decision making (leadership), and foreign alliances and partnerships. All of this requires direct and indirect military, and nonmilitary, lethal and nonlethal, and a mix of some or all of the above kinds of actions. In turn, this requires rethinking and redefining concepts that Larteguy points out in *The Centurians* (politics, propaganda, power, and victory). Lastly, this takes us back to where we began: all of this is dependent on end-state planning and an integrated strategic implementing process.

This is not to say that armed force cannot or should not be used, and used effectively to achieve political purpose. One only has to see how effective a few armed men with light weapons can be; one can also readily see how hard it is to defeat them and keep them from advancing their own political agenda by force. Force does have utility. But as General Rupert Smith warns us, "War no longer exists. Confrontation, conflict, and combat undoubtedly exist all around the world. . . .

Nonetheless, war as cognitively known to most noncombatants, as a battle in a field between men and machinery or as a massive deciding event in a dispute in international affairs—such war no longer exists. We are now engaged constantly and in many permutations, in war amongst the people. We must adapt our approaches and organize our institutions to this overwhelming reality if we are to triumph in the confrontations and conflicts that we face."[27] This is the indispensible lesson from the Algerian War.

THE SALVADORAN INSURGENCY

In the 1970s, chronic political, economic, and social problems created by a self-serving military-supported oligarchy began to generate another crisis in a long list of historical political crises in El Salvador. During that time, General Carlos Humberto Romero came to power with the support of those who thought he would be able to establish a regime strong enough to protect the interests of the oligarchy and to control the various forces agitating for reform. Yet by 1979, the situation was beyond control by repression.

The catalyst that ignited the violence in El Salvador was the military coup of October 1979 that ousted Romero as the last protector of the interests of the oligarchy. After Romero, the history of the country breaks into four clearly defined periods. The period immediately after the coup was one of almost complete disarray. None of the three major actors in the conflict—the military, the insurgents, and the United States—was ready for the aftermath of fifty years of authoritarian government. Then from the end of 1981 to the end of 1984, the Salvadoran revolutionaries seemed to unify and appeared to be well on their way to a military victory and the assumption of power in their own right. Clearly, the insurgents were ascendant. By the end of 1984, however, the Salvadoran armed forces had taken the best the insurgents could give and were beginning to regain control of the military situation. The period 1985–89 was a time of relative impasse and negotiations. The resultant 1992 Accords finally gave the Salvadoran government the objectives that it had articulated as early as 1984. The accords also allowed the demobilized revolutionaries to be assimilated into the liberal-democratic process. Interestingly and importantly, seventeen years later, former FMLN insurgents were democratically elected to the highest offices in El Salvador.

Disarray: A Look at the Antagonists, 1979–1981

In late 1979, the insurgents initiated a series of indirect and direct attacks throughout the country. The first form of attack was an intensive psychological campaign to challenge the legitimacy of the Romero regime and the junta that succeeded him. Then the insurgents attempted a "final offensive" in January 1981. They expected, as Che Guevara had taught, to gain a quick and total military victory. The effort was unsuccessful and was rationalized as the beginning of a "general offensive" that would ultimately lead to the final objective—to bring down the Salvadoran government.[28]

Some of the Salvadoran military saw the forces of change moving out of control at about the same time as the "final offensive." Moreover, they saw and understood what had happened to General Somoza's National Guard at the hands of the Nicaraguan people. The parallels were hard to ignore, leading to a general conclusion that the Salvadoran armed forces would suffer similar consequences if they did not act quickly to put out the fire of revolution in their own country.[29] To save the military institution (and themselves), the officer corps of the armed forces would have to break the alliance with the oligarchy and realign with political forces that could win popular support.[30]

These reformist officers were buffeted by both the left and the right, and also internally within the armed forces. The right opposed them because the officers were tied to the oligarchic interests that were threatened by proposed reforms.[31] They were labeled as traitors by the left because they were co-opting the political-social rationale of the Marxist-Leninist movement. That rationale "tracked" within a context in which the civil-military elites could not possibly make "correct" interpretations of and craft solutions to the country's problems; in those terms, the "vanguard of the proletariat" would have to be the ones to make those interpretations and determine proper solutions.[32]

Internally, within the armed forces themselves, the need to implement fundamental reforms, coupled with the struggle to establish a government based on a nontraditional, noncorporate model, created confusion and fragmentation.[33]

In the time-honored tradition of Latin American politics, the dominant military leadership established a civil-military junta. Their effort focused on sharing power with as many of the key power centers in the country as were willing to cooperate in an attempt to establish a unified control of the situation against the militant left. This network was to provide the basis for the subsequent organization to be established and empowered to effectively pursue the political-military dimensions of the struggle. Putting the concept of unity of command (and civil-military effort) into effect would also help ensure that all civil-military activity would be concentrated on the ultimate goal—survival.[34] The difficulty of achieving this goal when faced with an enemy specifically attempting to fragment and subvert a society can be, in essence, an organizational ("turf") war.[35]

For the Salvadoran reformers, the United States was the only source of external support that could make a difference. Yet during this period of disarray, the United States was also apparently confused. This problem was highlighted by the unwillingness or inability of senior policy makers in Washington to develop any kind of coordinated effort to deal with the situation in El Salvador—despite the general willingness and commitment of both the Carter and Reagan administrations to help.[36] A perceived "too little, too late" conundrum during this crucial period is but one example of U.S. confusion. Dr. Alvaro Magana, who acted as the Salvadoran president during 1982–84, argued that there appeared to have been no agreed-on, coherent strategy to achieve objectives and, indeed, no agreement as to what those objectives might be. Decisions concerning the allocation of "North American" resources to El Salvador appeared to have been made on the basis of what the minimum effort was that could be made while maintaining congressional support for administration policy. As a matter of fact, the only alternative policies examined involved different force levels for the Salvadoran army and specific amounts of economic and military aid.[37] Thus, issues addressed and decisions made were always tactical and short term in nature—the typical bureaucratic "in-box drill" of finding a "quick fix," selling it, and getting rid of the immediate problem.

For the insurgents' part, despite fifteen or more years of preparatory work and a decision to try to take control of the country, the revolutionary movement was not ready to control or to take advantage of the near anarchy of the time. The five various revolutionary factions that made up the FMLN had not yet unified in any significant way. Thus, to both revolutionaries and incumbents, the strategic solution to the mutual problems of confusion and disarray was, respectively, to create a real unity of command and effort. Both entities recognized that without a body at the highest level that could establish, enforce, prioritize, and continually refine cogent objectives, authority would be fragmented and there would be no way to resolve the myriad problems endemic to war and survival. That could mean failure or, at best, no win for either side.[38] Again, only the United States had the luxury of ignoring the central strategic reality of the period—the need for an organizational structure with the authority to plan and implement a holistic political-psychological-military counterinsurgency assistance effort.

The Period of Insurgent Ascendancy, 1981–1984

The leadership of the Popular Liberation Forces (FPL), the largest of the insurgent groups within the FMLN structure at the time, understood the importance of moral power in the strategy of conflict. They understood and were also responsive to the need to operationalize the classical principle of unity of command in war by engaging in the organizational war. However, the more militaristic leadership of the People's Revolutionary Army (ERP), flush with the insurrectionist victory of the Sandinistas in Nicaragua and supported by a strong push from the Cubans, prevailed. The ERP elected to do two things. First, they insisted on maintaining the five separate armed elements and gave the FMLN organization only umbrella status. Then, they determined to pursue a quick military victory over what was perceived to be a completely incompetent enemy.[39]

The first attempt at a quick revolutionary victory was launched in January 1981. Despite the failure of what was called the "final offensive," the FMLN managed to attain sufficient organizational unity, manpower, arms, sanctuaries, and outside support to generate a more or less continuous and growing military effort that explained the final offensive as the beginning of a "general offensive." That offensive lasted from the end of 1981 to the end of 1984. During that period, the guerrillas were able to organize, train, and logistically support units that were capable of mounting attacks with as many as six hundred men at virtually any time. During that period, they were also capable of controlling large portions of the national territory. Given that level of success, the revolutionary leadership argued that the general offensive would indeed lead to the achievement of the final objective of taking control of the Salvadoran government.[40]

This degree of military capability can be explained in terms of the exceptional external support enjoyed by the militant left. Indeed, the external support enjoyed by FMLN can also be explained by the failure to understand, support, or engage appropriate efforts on the part of either the Salvadoran government or its North American ally in the war against external support. Rather than actively attempt to counter the flow of arms and material, the Salvadoran security organizations and their mentors from the United States concentrated their efforts on finding a "smoking gun" that would clearly and legally implicate Nicaragua, Cuba, North Vietnam, and the Soviet Union in the support of the Salvadoran insurgents. They appeared to concentrate on building some sort of court case, ignoring or failing to take those actions that could counter and interdict the flow of external support.[41]

Despite the employment of various sophisticated and costly "platforms" designed to detect possible means and routes of entry into El Salvador, the effort never did establish the credibility of the "smoking gun" argument. Moreover, by ignoring or refusing to engage the insurgents' sources of external support (and the cross-border sanctuaries), the United States and the Salvadoran government provided the FMLN with their own protected "Ho Chi Minh Trail." As a result, the flow of support to the FMLN from abroad was never seriously impaired.

According to the leader of the ERP, Joaquin Villalobos, this trail was as important to the FMLN as the original trail was to the victory of the North in the Vietnamese struggle.[42]

While the revolutionaries were concentrating their efforts on the tactical-operational military aspects of the war, and the Salvadoran government was ignoring the strategic-level problem of external support to the insurgents, the government decided to concentrate its efforts in a struggle for the "hearts and minds" of the people. This was a struggle to gain legitimacy—and, thus, internal (Salvadoran) and external (U.S.) support. This was to be the first priority.[43] As a consequence, one of the first things the civil-military junta did on taking control of the government after the 1979 coup was to announce land, banking, and commodity export reforms. Subsequently, other reforms were promulgated—not the least of which were popular elections that really mattered. The degree of success these and other reforms may or may not have achieved is moot. The fact is that enough Salvadoran people were sufficiently convinced of the legitimate intent of the reforms that they did not support the insurgent cause to anywhere near the extent that might have been expected. At the same time, enough decision makers and policy makers in the United States were also sufficiently convinced of the value of the Salvadoran reforms that they continued their military and economic aid to El Salvador.[44]

The armed forces' leadership responded to the legitimization process on at least two fundamental levels. The military broke with its traditional right-wing allies and joined with moderate civilian politicians in an alliance to support the democratic constitutional process. The military went to extreme lengths to provide security for free elections and consistently demonstrated loyalty to civilian institutions—particularly to the office of the presidency. In the opinion of General Fred F. Woerner (a former U.S. Southern Command commander), that was probably the most significant reform of the decade.[45]

The Salvadoran military leadership also understood that guerrilla war had to be fought on diverse fronts and that soldiers and officers had to do more than shoot people to win the long-term struggle. Consequently, they took the necessary time and used the necessary resources to begin to change an 11,000–14,000-man "Praetorian Guard," accustomed to abusing its authority, into a more professional 50,000–55,000-man organization that could engage an enemy force without alienating the general citizenry. As far as the controlling military elite was concerned at that time, the central strategic effort could not be directed against a specific piece of territory or the enemy force. The central strategic effort was to be the basic legitimizing underpinnings of the Salvadoran government itself.[46]

As the armed forces began the process of reform and professionalization, they also developed the ability to fight a relatively intense conventional guerrilla war. The role of the United States was a positive one in these terms. However, during the 1981–84 period of insurgent ascendancy, U.S. assistance generally left much to be desired. At least a few senior decision makers were not particularly concerned, assuming that once the U.S. government had shown that it was prepared to continue to provide some help, the revolutionary movement would see the inevitably of defeat and simply go away. What this assumption did not take into account was the ideological commitment of the FMLN hardcore membership, the FMLN's strong belief that a prolonged war would cause the U.S. Congress to withdraw support for the Salvadoran government, and the strategic importance of the Nicaraguan-Cuban-Vietnamese-Soviet connection. In short, as far as the United States was concerned, there was no attempt to take the enemy into consideration.[47]

In summary, in this seemingly dark period in the history of the conflict in El Salvador, three things stand out in strategic perspective. First, legitimacy was reaffirmed as the factor that in the long term would prove to be more decisive than traditional military actions. In contrast, the *commandantes* (commanders) of the FMLN chose to all but ignore the counsel so generously provided by Mao Tse Tung, Vo Nguyen Giap, and their own "politicos" regarding the absolute need to supplement military action with a rigorous application of a moral dimension to contemporary war. Rather, they accepted Che Guevara's old argument that military action in revolution would create its own legitimacy.[48]

Second, the classical principles of unity of command and total effort were reaffirmed in the obverse. Both sides organized only to the extent necessary for survival and perhaps even for moderate success but not to the degree required to win. None of the principals to the conflict was able to overcome individual issues of turf, distrust, or lethargy to the degree necessary to develop an organization with the requisite authority to coordinate and implement a winning set of strategic political-psychological-military objectives.[49]

The third item of strategic significance was the fact that there appears to have been little cognizance of the importance of external support. Outside political, economic, and military aid to the insurgents made the guerrilla ascendancy possible. U.S. aid to the Salvadoran government probably saved that government. The Nicaraguan-Cuban-Vietnamese-Soviet nexus worked hard in the U.S. Congress and with U.S. public opinion to cut U.S. aid to El Salvador but did not quite succeed. Neither the United States nor the Salvadoran government effectively addressed the external sources of insurgent support. There was some effort to interdict the material assistance being provided to the insurgents from abroad. But what gave the FMLN the physical strength and psychological support it enjoyed was not the assistance itself or the routes that that assistance might have taken to get to the battlefield; the center of gravity in this context would be the *source* and the *consistency* of the support that was provided.[50]

The War Changes Direction and Goes to an Impasse

The strategy taught by Che Guevara, "There is only one road to victory, that of the armed struggle and the use of the peoples' methods of combat," remained in effect until mid-1984.[51] During that time, the FMLN left the cities, retired to the countryside, and began to mount major conventional-type attacks on the Salvadoran military. The FMLN gained control of large portions of the national territory, and France and Mexico granted diplomatic recognition to the insurgents.[52] However, by the end of 1984 and the beginning of 1985, the Salvadoran armed forces had acquired more manpower and resources, along with better training and equipment, than the FMLN. With these advantages and the legitimizing reforms, the government and the military began to reverse the tide of the conflict.[53]

By mid-1985, the insurgent leadership apparently agreed that a shift of the center of gravity had taken place. As Vietnamese general Nguyen Giap had taught, the shift was from the enemy military force to the source of the enemy's power.[54] Thus, the FMLN strategy became one of taking a relatively low military profile and focusing on nonmilitary centers of gravity—the legitimacy of the incumbent Salvadoran government and U.S. support for that regime.

To discredit the Salvadoran regime and disrupt the community of interest between the Salvadoran and U.S. governments, efforts centered first on the inability to provide real reform. As examples, the FMLN information war continually asserted that agrarian reform was not implemented and was a failure in any case; banking reform was a joke benefiting only the government; export reforms were irrelevant; elections were fraudulent; corruption of civil and military functionaries was widespread; and human rights were a sham. Second, on the diplomatic front, the FMLN and its external supporters continued the information war at the international level and worked hard to be perceived as the only entity that really wanted peace and reform.[55]

Yet given their fundamental "armed revolution" foco orientation, the FMLN broke down into small units and continued assassinations, kidnappings, and general terrorism on a carefully measured scale designed to constantly harass and intimidate the population and the government. These tactics were aimed at lessening regime legitimacy in terms of popular perception of its ability to govern and protect the citizenry. In this connection, the insurgents continued to attack transportation and communications networks and the general economic infrastructure of the country. The intent was to subvert government attempts to do anything that might improve the internal economy and the economic component of legitimacy. These tactics were also intended to impress on the United States the futility of its economic and military aid to El Salvador.

In addition, the FMLN mounted occasional spectacular attacks designed to give the impression that it still had good and relatively strong formations capable of military victory. As a consequence, insurgent violence, at whatever level, was not primarily military. Any military operation had political and psychological objectives as first priorities. Military objectives were tertiary. Thus, the threat in El Salvador—as in any insurgency—was multidimensional. In addition to the guerrilla war, at least five other wars were waged simultaneously: a war for perceived legitimacy; a "turf war" to achieve a complete unity of effort; a general war of information; a war to reduce outside support to the government; and a war of subversion.[56]

The Endgame

The endgame of the Salvadoran insurgency began in 1989 with serious negotiations between the government and the FMLN that were interrupted by a badly timed and ill-conceived final "final offensive." This military action cemented the U.S.-Salvadoran alliance and convinced large numbers of Salvadorans, as well as the international community, that the FMLN was only using the negotiations and the Central American peace process for the purposes of appearing legitimate and to gain time. About the same time, the Soviet Union began to disintegrate, and Soviet support for Cuba, Nicaragua, and the Salvadoran FMLN was withdrawn. It was only a matter of time before the insurgents would be completely unable to maintain their manpower and operations. Consequently, in 1992, a settlement was finally negotiated, and the war was ended.[57]

That settlement confirmed the general terms set forth by the government in 1984 and again by the Duarte government in 1987. Nevertheless, the FMLN was far more successful than it seemed at the time. The peace accords allowed the insurgents to integrate themselves into the constitutional electoral process, and seventeen years later, former FMLN insurgents were democratically elected to the highest political offices in El Salvador. That was the end of the idea that all revolutions came only from the "barrel of a gun."

It would be a terrible mistake to assume there is nothing to be learned from a common ordinary armed revolution such as that conducted in El Salvador over twenty years ago. To the contrary, the lessons learned from the Salvadoran, Uruguayan,[58] and other insurgency experiences around the world are all too relevant. Most important, these cases illustrate that there is a far superior alternative to violent and totalitarian models for achieving fundamental political change.

Key Points and Lessons

- In the beginning of the El Salvadoran insurgency, the FMLN leadership adopted a model of guerrilla war made popular by Che Guevara. He ignored the wisdom of Lenin regarding the

conditions necessary to conduct a successful revolution. He argued that an insurrection, in itself, would create the conditions necessary to generate a revolution. Such an insurgency could be organized by a relatively small but mobile group of guerrilla fighters.
- During the period between 1985 and 1989, in addition to the more or less conventional guerrilla war, Salvadoran civil-military leaders and FMLN leaders came to understand lessons from the earlier Vietnamese experience in Southeast Asia. That is, there are wars within the war. In addition to the guerrilla war, at least five other wars or combinations of wars were waged simultaneously: a war for relative perceived legitimacy; a "turf war" to develop the organization necessary to achieve a unity of effort; a general war of information; a war to reduce external support to the opponent; and a war of subversion.
- Thus, the Salvadoran government and FMLN strategies were altered from time to time in recognition of changing political-military conditions, but both parties to the conflict relied primarily on a strong military component in their efforts to "win." In the end, after a negotiated settlement in 1992, the government achieved the goals it put forth in 1984 and again in 1987. At the same time, the insurgents were incorporated into the democratic political process. That was the validation of the idea that there is a viable alternative to violent models for achieving fundamental political change.

Sun Tzu reminds us that "to win one hundred victories in one hundred battles is not the acme of [political-military] skill. To subdue the enemy without fighting is the acme of skill."[59]

CHAPTER 2

New "Kindler and Gentler" Revolutionary Lessons from Peru

The Resurgence of Sendero Luminoso

The 2008 and 2010 *Latinobarómetro* polls, taken in eighteen countries of Latin America, underline the fact that even though most Latin American countries' gross domestic product (GDP) has been improving since 2001, there are deep flaws in democratic political systems throughout the region. The relative popular dissatisfaction stems from deep-rooted socioeconomic inequalities, distrust, and lack of confidence in the police, the legislature, and the political parties. There are also rising popular expectations along with growing popular consciousness of nonexistent rights. Peruvians are particularly disgruntled. Peru's economy has grown faster than any other of Latin America's bigger countries; yet of all the countries polled, Peru demonstrates the greatest dissatisfaction. Fewer than 30 percent of Peruvians are satisfied with the way democracy works in their country. As a consequence, more than 70 percent of the population would agree that Peru "is in serious shape."[1] Peru, now—as it did in the 1960s—appears to be an insurgent's dream.

Thus, the resurgence of the Sendero Luminoso (Shining Path) insurgency continues to evolve. Accordingly, this chapter illustrates what Jorge Verstrynge calls the "revalidation of guerrilla warfare."[2] A translation of that Marxist-Leninist rhetoric dictates that concepts of traditional guerrilla warfare are being superseded by those of Sun Tzu's "indirect war," Clausewitz's "war by other means," and V. I. Lenin's "war by all means."[3] Importantly, the term "revalidation" now dictates that the insurgency phenomenon must move from Che Guevara's violent military-oriented *(foco)* approach to compel rapid radical change, and revert to Lenin's, Mao Tse Tung's (or Mao Zedong's), and—now—Verstrynge's softer and more subtle use of multidimensional combinations of propaganda, corruption, subversion, coercion, and time to achieve indirectly the kind of power that creates its own legitimacy and generates radical political change.[4]

These actions and the actors who perpetrate them have the capability to seriously threaten the stability, well-being, and interests of individual nations and the global community. Peru and the resurgence of Sendero Luminoso is a case in point. A fundamental empirical challenge is to identify salient changes and continuities in this particular insurgency movement and thereby contribute to the process of rethinking the general insurgency phenomenon and shaping a strategic response.

THE POLITICAL-HISTORICAL BACKGROUND OF THE PERUVIAN INSURGENCY SITUATION

The political history of Latin America shows that of all the nation-states carved out of Spain's American empire, Peru has clung most tenaciously to the traditions of the Iberian motherland. In Peru, there was no popular clamor for independence. Independence came through the military intervention of General José de San Martín (from Argentina) and General Simón Bolívar (from Venezuela). The mantle of colonial viceroys, captains-general, and royal judges fell on the shoulders of Spanish landholders, bishops, civilian *caudillos* (strongmen), and military officers. The general political condition in Peru passed from King Ferdinand VII into the hands of General don Simón de Bolívar. And despite the promulgation of a U.S.-style constitution, Peruvian national leaders, from Bolívar to those in power until nearly the end of the twentieth century, were chosen by the oligarchs, the clergy, and the army. Serious disagreements within the political process were usually resolved by "gunfire and *coup*," rather than by compromise or elections.[5] One of Peru's first reformers, Manuel González Prada (1848–1918), argued that "the form of the country's government should be called an extension of the [Spanish] Conquest and Viceroyalty."[6]

Prada's most distinguished disciples included José Carlos Mariátegui and Víctor Raúl Haya de la Torre. They borrowed ideas from Karl Marx, V. I. Lenin, the Mexican Revolution of 1910–20, and pre–World War II European socialism. They called for nationalization of land and industry, a united workers' and intellectuals' political party, a diminution of the property and power of the Catholic Church, and the unity of Indo-America against the economic and political imperialism of the United States. Meaningful and substantive reform, however, did not come until after the election of President Fernando Belaunde Terry. In 1963, he and his more moderate followers successfully opposed those elites who had represented "Viceregal Peru" over the preceding four centuries.[7]

The genius of Belaunde was illustrated by the fact that he was the first Peruvian politician to think in terms of a plural society—not central rule by the elite "40 families of Peru" or the idealized rule of the Indian majority of the country. He wanted to integrate the entire Peruvian society. He also thought in terms of the whole of physical-geographical Peru—on both sides of the Andes Mountains—not just the narrow coastal plain and the great city of Lima, but the entire eastern mountain slopes, valleys, rivers, and jungles of the rest of the country. Belaunde did not accomplish all his hopes and dreams, but he did manage to get a comprehensive Agrarian Reform Law passed by the Congress in 1964 and subsequently promulgated over five hundred legislative acts designed to stimulate education, housing, social security, and local rule. This reform process, however, was never completed. Consequently, Peru is described as having remained the most obstinately feudal society in all the Americas to this day.[8]

A deeper understanding of the historical-political background within which the Sendero Luminoso movement is operating must begin with two controlling realities. First, Abimael Guzmán's vision of the "democratic liberation of Peru" is key to understanding the contemporary situation. It performs the universal Marxist-Leninist function of providing the ultimate political objective, the controlling mechanisms, and the basic strategic plan for the movement. Second, as a corollary, a look at Sendero's action program from 1962 to 1992 is instructive. These contextual realities begin the process of explaining Shining Path's challenge to Peru, along with illustrating the brilliance of a highly structured insurgency model.

GUZMÁN'S VISION FOR PERU

Abimael Guzmán identified the origins of the Sendero Luminoso insurgency in Peru and defined the nation's central strategic problem as the lack of legitimacy of all Peruvian governments since the Spanish conquest.[9] He further identified the primary objective of the insurgency as power. Power is generated by an intelligent, well-motivated, and highly disciplined organization with a vision and a purposeful, long-term program for gaining control of a state or a society. Power is maintained and enhanced as that organization gradually replaces the state.[10]

The general vision of the Sendero Luminoso organization is to destroy the old foreign-dominated political system in Peru, to take power, and to create a "nationalistic," "Indian," and "popular" democracy. Guzmán argued that the original basis for Peruvian socialism is in the pre-Columbian Indian (Quechua) community. The revolutionary challenge is rooted in the concept that the incumbent governmental system is not doing what is right for the people and that the Sendero political philosophy and leadership will. As an example, Sendero has acted as the de facto government in this power vacuum. It has provided significant public services that had been lacking when the Peruvian government was supposedly in control. Additionally, Sendero Luminoso established legal codes that, while considered hard, were acknowledged to be fairer than the corrupt governance administered by the government officials that had been in control in the past.[11] Thus, Guzmán considers himself to be initiating a third epoch in the history of Peru, one that will reestablish "true democracy."[12] The Sendero Luminoso concept of true democracy (absolute political freedom) is a classical Marxist-Leninist dialectical leap of logic: "Absolute freedom is the right of the strongest to dominate."[13]

Abimael Guzmán's first and continuing concern, however, centers on organization. The preparatory activities for achieving his vision were to establish a dedicated cadre for a revolutionary party, a guerrilla army, and a support mechanism for the entire organization. This fifteen- to twenty-year effort would lay the foundations for the subsequent long-term struggle (called the long war, or *guerra prolongada*) and ultimate victory. Organizational health, breadth, and depth, then, provide the bases for the primary Sendero Luminoso measures of effectiveness. The organization, not its operations, would be the key to success.[14]

Thus, Sendero Luminoso is not an "organization of the masses." At the start of the "people's war" in Peru, in 1980, Sendero was estimated to be numbered at only 189 militants.[15] Even by the time Guzmán was captured in 1992, the number of Shining Path fighters was estimated to be well below 10,000.[16] In 2009, the U.S. Department of State estimated Sendero membership to be only 300–500.[17] Larger numbers were not considered necessary. In "ungoverned territory" or poorly governed territory, such as that on the eastern slopes of the Peruvian Andes, only a few armed personnel are required to establish Sendero's writ. As a consequence, Sendero Luminoso has not developed spontaneously as a result of societal causes producing activists; rather, it is a movement in which activists are taking advantage of social causes. In short, Shining Path is the Leninist "vanguard of the proletariat."

In that connection, the scientific (Marxist) name for conducting defensive war against a bourgeois enemy is "dictatorship of the proletariat." The effort that breaks up and finally defeats an opposing state and its bourgeois internal and external accomplices is conducted by the "vanguard of the proletariat." Sendero leadership, through the period 1962–92, was convinced that the defeat of the bourgeois enemy would be accomplished primarily by an organization that could intimidate the public as well as the government with carefully applied and brutal military force (again, the dictatorship of the proletariat).[18]

Guzmán's Action Program for the Democratic Liberation of Peru, 1962–1992

As early as 1962, thinking that Peru was demonstrating the classic Leninist symptoms of a revolutionary situation,[19] Guzmán began planning a long-term, multistage, sometimes overlapping program for reestablishing a people's democracy.[20] The plan was based on Lenin's and Mao's multistage models for conducting revolutionary war but was heavily influenced by Che Guevara's violent, relatively short-term approach to revolution.[21] Thus, using his own doctrine and logic, Guzmán originally envisioned a five-stage program that could be altered as conditions might dictate. Accordingly, he added a sixth stage to his revolutionary program at the time of his capture and jailing.[22] The original five stages of the program were the following:

Stage 1: The Organizational Stage, 1962–80. The essential first efforts began in 1962, taking the time necessary to lay the organizational foundations for the subsequent "armed struggle." During the 1960s and 1970s, Guzmán concentrated on doctrine and leadership development and on expanding his organization's relationships with relatively isolated peasant communities in the outlying districts in the highlands around the city of Ayacucho. Then, in 1978, Sendero completely disappeared from public view.[23]

Stage 2: Moving into the Offensive, 1980–82. As Peru was returning to civilian rule after twelve years of military government, Guzmán moved from the organizational phase of revolutionary development to the offensive and began to attack the symbols of the bourgeois state. Sendero bombed public buildings and private companies, hanged dogs and cats from lampposts as warnings to functionaries and supporters of the illegitimate state, and initiated a series of attacks on and assassinations of local public figures. The objectives of this violence were to "attack the glue that holds society together," to destroy direct communications between the government and the population, and to begin to create a political vacuum that would allow Shining Path to become the de facto authority in the areas uncontrolled or abandoned by the state.

Stage 3: The Generalization of Violence, 1982–83. Essentially, this stage was a more violent continuation of stage 2. It began in March 1982 with a major attack on the Ayacucho Department prison and the Robin Hood–like release of all its prisoners. This operation was followed by another spectacular event in December. Sendero attacked Lima's electrical grid, destroyed four high-tension towers, and caused a complete blackout in the capital and six other cities. Minutes after all the lights were out and everyone was in the streets wondering what had happened, a huge hammer and sickle was lit on a hill overlooking the city of Lima in celebration of Guzmán's forty-eighth birthday.[24]

Despite these spectacular events, this phase of Guzmán's revolutionary war strategy was directed primarily at specific key individuals. A most effective tactic was the coordinated use of assassination and posted death threats to disrupt, paralyze, and eliminate local governing institutions. As an example, moving into an area, Sendero would declare the region to be a "zone of liberation." Large numbers of community leaders, administrators, and other "traitors" were rounded up and, after a public trial, were hanged, shot, or beaten—depending on the seriousness of their individual crimes against the revolution. After setting this example, Sendero would publish a "death list" of all those individuals remaining in the area who were slated to be brought to justice. Terrified, those whose names appeared on the death list left the community in the hands of Sendero Luminoso functionaries.

Stage 4: The Consolidation and Expansion of Political and Logistical Support Bases, 1983–92. This phase of the Sendero program was also known as the "programmatic isolation of the center," that is, the isolation of the city of Lima. From 1983 to the time of Guzmán's capture, Sendero escalated and de-escalated its intimidating "armed propaganda" efforts as the general strategic situation required but always worked to consolidate and expand its political and logistical support bases throughout the country. By the end of 1992, Shining Path had extended its presence into 114 provinces across all departments of the country, including the area that produced over 60 percent of the world's supply of coca leaf, the Huallaga Valley. This effort essentially left only the coastal departments and the large cities under central government control and provided Sendero a commanding position from which to envelop the city of Lima.

Such rapid territorial expansion would have been impossible without substantial financial support. Without traditional external aid from revolutionary-minded governments or rogue states, Sendero relied on extortion from businesses throughout Peru and on the extraction of "taxes" from the illegal drug trafficking industry. In 1992, Sendero Luminoso was estimated to make between $30 million and $100 million a year from Peruvian narco-traffickers.[25]

Stage 5: Besieging the Cities and Bringing About the Total Collapse of the State, 1989–92. In the original plan for taking control of Peru, this was to be the decisive and final phase of the revolution. But this stage of the revolutionary effort was not scheduled to take place until the interior support bases were consolidated, the leadership nucleus of the movement was sufficiently large and effectively prepared to administer the state, and the major population centers were either strangled economically or subverted psychologically to the point where a relatively small, but powerful, military assault could bring about the desired result.[26]

Summary

In short, throughout the period 1962–92, coerced radical change was the essential nature of the threat from Shining Path.[27] Then, after 1992, the minimum objective of the sixth stage of the revolution was to continue to make a bad situation in Peru worse and to continue fifth-stage preparations to take ultimate control of the state. Guzmán and Sendero leadership are not known for dwelling on minimum objectives, however. Their ambitions, goals, and dreams go well beyond making a bad situation worse.[28]

GUZMÁN'S NEW SIXTH STAGE OF THE REVOLUTION, AND THE ROAD (SHINING PATH) AHEAD

At the time of Abimael Guzmán's capture and jailing in 1992, Peru's President Albert Fujimori declared that the Shining Path had been defeated and that terrorism had been eradicated from Peru.[29] Virtually everyone in Peru, the Andean region of South America, and all the wishful thinkers in the global community expected, hoped, and prayed that Sendero Luminoso was indeed defunct. Nevertheless, after a few years of "remaining relatively quiet," Sendero began to reemerge. Incidents involving the Shining Path increased yearly, from about 200 in 2002 to 426 in 2005 to more than double that in 2008.[30] Wishful thinkers everywhere—except in the jungles (the Peruvian *selva* on the eastern slopes of the Andes where the waters drain into the Amazon Basin)—hoped that Peru's interior minister at the time, Octavio Salazar, was right when he said, "Sendero Luminoso is a threat to public safety in the jungles, [but] no longer poses a threat to [Peru]."[31]

However, reality challenges wishful thinking. Abundant and clear evidence exists that since 1992–93 Shining Path has morphed from a dormant insurgency to being "defenders of coca growers," to building a new base of support, to resuming a serious "armed propaganda" campaign.[32] Moreover, the evidence demonstrates that Sendero Luminoso is promulgating a more benign ideology (Verstrynge's multidimensional "revalidation of guerrilla war") closely aligned with Lenin's political-psychological-agitation concepts and Mao's notion of "protracted war."[33] Thus, probably the best place to begin to look for clues regarding Shining Path's current and future direction would be the "new" and additional sixth stage of Guzmán's program for conducting contemporary revolutionary war.[34]

Stage Six of Guzmán's Program for the Liberation of Peru: Preparing for the Total Collapse of the State and for the World Revolution

A closer look at Guzmán's sixth stage of the Peruvian Revolution indicates that, like a good football coach when a game has gone awry, he takes his team back to fundamentals. Guzmán reminds us that the primary objective of the revolution is power. "All else is illusion."[35] Thus, in reality, the sixth stage of Guzmán's variation on revolutionary war is the first and, doctrinally, an extremely long-term stage of a renewed national, regional, and global effort to achieve power. In these terms, a successful revolutionary effort cannot be constrained by time, methods, or limited objectives. All the time necessary to achieve the ultimate objective of taking control of a targeted government must be taken. No shortcuts will work. All the various methods of conducting war (conflict) to achieve victory must be used in innovative combinations. Military effort alone will not work. And if the vanguard of the proletariat has to make unpleasant (temporary) deals with its bourgeois adversaries and debourgeoised allies, so be it. The ultimate objective of compelling fundamental political-economic-social change and taking complete control of a targeted state or region must not be compromised in any way.[36]

Thus, the direction Shining Path appears to be taking in the first part of the twenty-first century has been articulated in Jorge Verstrynge's "revalidation of guerrilla warfare," within the context of his notion of peripheral (indirect) war.[37] Abimael Guzmán's theoretical application of that guidance can be examined in the context of the following five elements: (1) the reorganization of the vanguard of the proletariat, (2) the need for a broad political front as the basis from which to pursue radical political change, (3) a continental scope of effort, (4) a new organization and role for the People's Liberation Army (PLA), and (5) a move toward world revolution.

Toward the Creation of a New Vanguard of the Proletariat. As noted above, Guzmán's first and continuing concern remains centered on organization. The preparatory activities toward achieving his vision were and continue to be to establish a dedicated cadre and a revolutionary party, a guerrilla army (the PLA), and a support mechanism for the entire organization. These long-term efforts are intended to lay the foundations for the subsequent long-term struggle (prolonged war) and ultimate takeover of the Peruvian state. Again, the organization—not the operations—is the key to success.[38]

In sum, Sendero Luminoso appears to be in a rebuilding process, having returned to stage 1 of Guzmán's original revolutionary plan. In this stage, the revolutionary leadership must again concentrate on doctrine and leadership development, expand the organization's relationships with other regional and global political movements, and create a receptive political-psychological environment in Peru for the revolutionary movement. In that connection, according to another of Guzmán's more recent mentors, Abraham Guillen, a more specific requirement is to "[c]reate a popular front of a combination of

Christians, Socialists, trade unionists, intellectuals, students, peasants, and the debourgeoised middle class who will march together to defeat sepoyan (regional) militarism and U.S. imperialism."[39]

Toward Mass Support for Radical Change. North Vietnam's General Vo Nguyen Giap explained one of the principal reasons for the success of the North Vietnamese people's revolution: "The Vietnamese people's war of liberation was victorious because we had a wide and firm National United Front, comprising all the revolutionary classes, all the nationalities living on Vietnamese soil, all the patriots. This Front was based on the alliance between workers and peasants, under the leadership of the party."[40] This kind of political effort might conceivably have to be a part of a legal liberal-democratic process; early in the revolutionary struggle, however, Guzmán did not consider that idea worth serious consideration, because he did not trust the liberal-democratic process or the voters to do the "right thing." He believed that only a brutally violent Sendero Luminoso could propel Peru toward "true democracy."[41]

However, after his capture and jailing, Guzmán must have asked himself the question "If too many accidents demonstrate the same phenomenon, can you say they are accidents?" The answer was clear: "No, at this moment, you have to say that there is a rule here. . . . But, the rule should not be understood literally in a narrow manner. . . . The key is to grasp the essence and apply the principle."[42] Lenin, Mao, and a majority of other successful insurgent leaders and thinkers have understood this. Thus, Guzmán and Sendero's leadership have come to understand General Giap's dictum, and now it is part of the sixth stage of Sendero's program for the liberation of Peru.[43] Politics is, indeed, a continuation of war by other means.

Nevertheless, the new Sendero Luminoso will likely find the task of developing a united front political organization difficult and painful. In contrast with past strategies, they will have to appreciate the importance of moral as well as de facto and de jure legitimacy. They will have to understand that the vanguard of the proletariat cannot bring about a successful revolution without the support of the people they purport to lead. They will have to rethink their ideas of tolerance, cooperation, equality, and compromise as they try to build a political coalition that could threaten to topple an incumbent government by employing democratic means. They will also have to organize a campaign of information gathering and then develop public support for a new, people-oriented Shining Path agenda based on this information. These are some of the strategic-level analytical commonalities that have proved, over the years and throughout the world, to generate success in the political conflict environment.[44]

Thus, Sendero leadership now reckons that a broad political front representing mass popular support will be the legitimizing strategic foundation for a true vanguard of the proletariat. As a first move in this direction, in 2010, Comrade Artemio (operating in the Huallaga region) called on the Peruvian government to enter into a dialogue. He is quoted as having said, "We want the people to understand that if we do them harm, we feel ashamed, and we ask forgiveness from the families we have cast into mourning." He went on to request talks with Peru's civil and military leaders in the presence of mediators.[45] Interestingly and importantly, Artemio's statement is completely in character with Lenin's advice regarding (temporary) concessions to the enemy in war, but its timing was a bit off. A highly competitive political campaign was being conducted at the time and a new government was elected in 2011. Although there are a few negative indicators, its position cannot yet be determined.[46]

Toward a Continental Revolution: The Sendero-Narco-FARC-Venezuelan Nexus. The current phase of the long-term revolution must be enhanced by the resources of brother states and nonstate political actors. These political actors would include but not be limited to organizations such as Colombia's FARC (Revolutionary Armed Forces of Colombia); Venezuela's activist popular militias, the Brigades (Las Brigadas Simon Bolivar); the Continental Coordinating Board (CCB, La Coordinadora Continental Bolivariana); and various narco-terrorist transnational criminal organizations (TCOs) operating in the Western Hemisphere. There has been a close relationship between Sendero Luminoso, the illegal transnational drug trafficking industry in Peru, the so-called narco-terrorists (FARC and AUC) in Colombia, and the Bolivarian opportunists in Venezuela and the rest of the Andean Ridge countries since 2002.[47] At present, President Ollanta Moises Humala admits that most of the coca grown in the country is being produced under the protection of Sendero Luminoso in five thousand square kilometers of the Apurimac and Ene River Valleys (VRAE) on the eastern slopes of the Andes. Additionally, the minister of defense, Jaime Thorne Leon, has stated that this production and protection is being accomplished with the participation of the Colombian FARC.[48]

The equation that links narcotics trafficking to insurgency turns on a combination of need, organizational infrastructure development, ability, and the availability of sophisticated communications and weaponry. For example, traffickers possess cash and lines of transportation and communication. Sendero Luminoso possesses soldiers, discipline, and organization. Traffickers need these assets to protect their products and transportation networks, as well as to ensure freedom of movement and action within their production, transportation, and marketing areas. The Shining Path revolutionary movement needs logistical and communications support—and money. Thus, Sendero leadership continues to see the Sendero-narco-FARC-Venezuelan nexus as an essential part of the general campaign to "prepare for the total collapse of the state, and for the World Revolution."[49]

Billions of dollars in profits coming from the United States, Europe, and elsewhere in the world; easy access to high-tech weapons and communications systems; and willing volunteers to serve as hired guns constitute much of the economic portion of the power equation. The political portion of the power equation is manifested by the ability to buy influence, assassinate opponents, and thereby control local and national political systems. Additionally, both groups possess relatively flat organizational structures that, when combined, can generate a more efficient and effective organization than any slow-moving bureaucratic, hierarchical governmental system. Deep pockets and flat organizational structure also mean that members of the alliance can move, shift, diversify, and promote operations inside and across national borders at will. That combined organizational advantage is a major source of power in itself.[50] At the same time, the various components of the alliance may be directly performing the tasks of government internally and externally, acting as sovereign states within a traditional nation-state.[51] In Verstrynge's opinion, this kind of alliance "constitutes the greatest threat that the world must confront now and in the future."[52]

The Creation of a People's Liberation Army: Toward a Prolonged War. Sendero's leaders know that they lack the conventional power to directly challenge the Peruvian government or the government of any of Peru's neighbors. As a consequence, the revolutionary leadership understands that irregular, asymmetric warfare is the only logical means through which to compel a stronger political entity to do one's will. As Abraham Guillen teaches, "In a war of liberation the final victory is not decided by arms, as in imperialist wars. . . . That side wins which endures longest: morally, politically, and economically. . . . If one knows how to employ strategically the factor of time and space with the support of the population. . . . The side that knows how to endure the longest will ultimately win."[53] In these terms, the leadership understands that war is no longer limited to using conventional guerrilla-military violence to bring about desired political-economic-social change. Rather, all means that can be brought to bear on a given situation must be used to bring indirect pressures on a targeted government to (1) facilitate and accelerate the processes of state failure, (2) generate progressively greater freedom of movement and action for Sendero's continued development, and (3) as a result, prepare the way to force a radical restructuring of Peru's government and economy.

This kind of irregular war—based on the notion that the human terrain is the main contemporary center of gravity—relies primarily on words, images, perceptions, and ideas. Moreover, the only ethics are those that contribute directly to the achievement of the ultimate political objective. The only rule is that there are no rules. Even so, the idea of unrestricted war cannot preclude direct or indirect military operations.[54] Theoretically, the formation of a People's Liberation Army is imperative, for such an army is an integral part of *la guerra periferica* (peripheral or indirect war) and the whole revolutionary process.[55] And, again, Guillen reminds us that experience teaches that "To win the support of the population, arms must be used directly [and obviously] on its behalf."[56]

Thus, from Guzmán's perspective, an integral part of "preparing for the total collapse of the state and for the world revolution" is the task to convert the brutally terroristic Popular Guerrilla Army of the past into a more substantial and conventional People's Liberation Army (PLA). The intent is to ensure that the overall political-military efforts of the PLA would be directed toward developing a more protective and benign relationship with the masses that would, in turn, enable the protraction of the *guerra prolongada* until ultimate political success is achieved.[57]

Toward World Revolution. Guzmán argued that the world revolution would be sustained over time through successive "cultural revolutions" until the peoples' values have been completely changed and the ultimate truly "communist phase" of history is established. That phase will have been recognized as having taken place when Peru joins together with other "social-democrats" and allies to form a new global or regional communist political organization. The seriousness of this effort for the "democratic liberation of Peru"—and the rest of Latin America—cannot be dismissed as too complex, too ambitious, too far into the future to deal with, or too unrealistic. The revolutionary dream is too powerful to reject or wish away. The revolutionary dream is powerful and alive in Latin America and much of the rest of the world. As a result, in the case of Peru and elsewhere, we must remember that only when "an unjust government collapses and new Socialist governance is imposed [will] the people . . . begin to enjoy the benefits of love, happiness, peace, and well-being."[58]

This dream provides major inspiration for would-be revolutionaries all over the world. Sendero's revalidation of Verstrynge's guerrilla (peripheral) war provides a highly developed architecture for a "kinder, gentler" and pragmatic approach to fundamental political-economic-social change. In *Dragonwars*, J. Bowyer Bell reminds us that "much of the world is ripe for those who wish to change history, revenge grievances, find security in a new [political-economic] structure, or protect old ways. Those individuals are not easily swayed for they seek not tangibles, but seek the realization of a dream, the rewards of history. These aspirations are not easily accommodated. [Thus], the next century, as the last, offers the prospect of war—old wars, but also new wars that deploy assassins, plagues, and the unconventional."[59]

Back to the Dictatorship of the Proletariat

This brings us back to the role of the vanguard and the dictatorship of the proletariat. Sendero Luminoso considers itself to be the organization that provides the vision, the plan, and the controlling (dictatorial) mechanisms for achieving regional power in Peru and continental power in the Western Hemisphere. That effort is accomplished by the systematic (dictatorial) application of indirect and direct, state and nonstate, military and nonmilitary, lethal and nonlethal actions, and a mix of some or all of these kinds of actions. That is, "Any war is merely the continuation of peace time politics by other means."[60] Thus, Guzmán applies Lenin's notion of *total war* to defensive war on three different levels. First, *all* ways and means—political-economic, social-psychological, and military-paramilitary—must be used to destabilize the bourgeois state and generate the erosion and ultimate defeat of liberal democracy and market capitalism; second, one must achieve total political control of government (radical change); and third, one must use a totality of ways and means to compel fundamental political change and new values.[61] Then and only then will the vanguard of the proletariat have made a permanent change in the political, economic, and social system of a targeted political entity—one piece at a time.[62]

BACK TO THE POLITICAL SITUATION IN PERU, AND SOME SOFT POWER RECOMMENDATIONS

As early as the 1980s, General Edgardo Mercado Jarrín and some of his military and civilian colleagues warned that if the Peruvian government and its security institutions (and international allies, if there were to be any) lose the fight against Sendero Luminoso, this would not be the result of large numbers of PLA soldiers overpowering the regime, the police, and the military. It would be because a handful of highly motivated and well-trained PLA soldiers were unmoved by traditional counterinsurgency, counterdrug, and counterterrorism tactics and conventional military-police means of deterrence. Most important, defeat by Sendero would be the consequence of the Peruvian people being unwilling to cooperate actively with and support their democratically elected government.[63]

This kind of war is won by altering, directly and indirectly, the political-psychological factors that govern the internal conflict situation. Directly, actions speak louder than words. In these terms, the principal effort of the government and its international allies must center on obvious and comprehensive political reform. Peru's institutions are ill equipped to deal with popular demands in a coherent manner, however legitimate they may be. Political parties have lost credibility, government agencies have little or no credibility or are simply absent, and security organizations often act more like invading foreign armies than institutions that are supposed to provide individual and collective security. Under these circumstances, piecemeal steps toward electoral reforms have proved to be destabilizing.[64]

Well-planned and holistic reform, then, would seem to be the ideal political objective of a government aiming toward the survival of the state. If the Peruvian people cannot see and experience meaningful change and reform, then demagogues, populists, warlords, drug lords, criminals, Sendero Luminoso, and other insurgent organizations will likely compete for control of the failing state. The result of that competition will be determined by the best-organized, best-motivated, best-disciplined, and best-armed political actor on the scene at the time. The logical conclusion would be a narco-criminal state, a populist direct democracy, or a new "People's Democratic Republic." Peru's long-term future appears to be bleak, indeed.

What, then, is to be done? Ambassador David Passage, who has a great deal of experience in global security matters, provides a beginning point from which to start the admittedly slow, unsure process toward comprehensive reform and political viability. He outlines the problem, the challenge, and the task at hand. The problem is straightforward: there is little or no significant or consistent engagement by the people on the side of the government. The challenge is equally blunt, for it is to get target audiences to think through the problem and answer the question "What will happen if Sendero Luminoso does gain control of the state, and do you really want to live with that possibility?" A corollary question is "What do Shining Path leaders mean when they say they want to establish a 'true democracy' in Peru?"

The task, of course, cannot be quickly or easily accomplished. The first part of the task is to find ways, in the new information era, of identifying critical issues that the government and the people face and to explain clearly and persuasively what the incumbent regime is doing about them. While these efforts are ongoing, the second part of the task is to convince the people to support the government actively as it works to achieve the desired political objectives.[65]

It is unrealistic to think that an outside power like the United States should or could initiate and maintain a viable liberal-democratic reform program in Peru. There are, however, some initiatives that might be helpful. Experience could remind us that social engineering projects are best undertaken by internal actors. In any event, there is no silver bullet. Experience should also remind us, according to Ambassador Passage, that the fundamental message to the people of Peru—or any other potential failed state—needs to be "This is your country; the kind of democracy it is going to be is up to you, not the United States, the United Nations, or any other international political actor. Are you willing to help restore order, and law, and civility; or are you going to sit quietly while those who seek to destroy the good things you have are preparing to build their own institutions?" Ambassador Passage argues that this message must, in the end, oblige the country to examine its problems, challenges, and tasks; think it all through; and then provide realistic solutions.[66]

The United States took this kind of democratization and reform approach in El Salvador in the early 1980s. First, the United States made clear that this was El Salvador's war—not America's. Salvadorans would win their war, or they—not the United States—would lose it. Second, the United States would help "professionalize" (train) El Salvador's armed forces to fight successfully against the enemy they faced. And third, as the price of U.S. assistance, the United States required the Salvadoran government to change its attitude and approach to its own people fundamentally to win their active support and cooperation.[67]

That approach worked in El Salvador, as it had for the communists in Vietnam's People's War, in Italy's war against the Red Brigades, and elsewhere in postcolonial Africa, Asia, and the Middle East.[68] In El Salvador, President José Napoleón Duarte pushed needed economic, social, and political reforms through the national legislature. He also built a substantial popular consensus in favor of his liberal-democratic government as a result of exemplary public diplomacy efforts. At the same time, the Salvadoran government and its armed forces, with minimum U.S. help, changed the military from a bunch of thugs into a competent fighting force willing to subordinate itself to the legitimate civil authority. Within less than ten years, the counterinsurgency war was won, and the U.S. mission was complete.[69]

Again, there are no silver bullets, and there are no guarantees. However, over the past fifteen to twenty years, the Sendero Luminoso leadership has come to this conclusion: "Even though every conflict situation differs in terms of history, geography, culture, and specific circumstances, there are analytical commonalities (strategic-level governing rules) that can be seen operating in every case. One of these rules—that of the decisiveness of popular support—must be incorporated into the long-term struggle to take control of Peru."[70] The associated liberal-democratic countertask of the Peruvian government is twofold. First, the political-military leadership must grasp the essence of that "rule" and apply the principle, that is, they must convince the population that they have embarked on a realistic program of fundamental popular reform. Second, the Peruvian leadership must learn for itself and help the rest of the country understand Shining Path's definition of "true democracy," for which French philosopher and novelist Albert Camus offers a succinct definition: "Absolute freedom [true democracy] is the right of the strongest [the dictatorship of the proletariat] to dominate.... Absolute justice is achieved by the suppression of all contradiction [democratic centralism]: therefore it destroys freedom and democracy.... The 'dialectical miracle' is the decision to call total servitude freedom or democracy."[71] Soft power, public diplomacy, and information operations campaigns must challenge target audiences to think through carefully this definition of "true democracy."

KEY POINTS AND LESSONS

- Peru is experiencing a resurgence of the supposedly defunct Sendero Luminoso. That reemergence of a violent military-dominated revolutionary philosophy was and remains deeply engrained in Guzmán's and Sendero thinking. The argument was and is that brutal organized violence is an expedient shortcut to compel radical political change.
- Nevertheless, past experience in Peru and elsewhere in the revolutionary world mandates moving to a "kinder, gentler" and longer-term use of combinations of political-psychological-coercive methods to generate a modern revolution (the "revalidation of guerrilla war").
- That internal dynamic in Peru is exacerbated by two additional and powerful external elements/players closely associated with Sendero Luminoso: the illegal drug trafficking industry; and the more subtle and politically oriented transnational Bolivarian (Venezuelan) Latin American unity effort.
- Each set of kinder, gentler—but still violent—nonstate actors involved in the direct and indirect subversion of the Peruvian political-economic-social system has its own specific and different motivation. The common denominator, however, is the political objective of compelling radical change in the Peruvian state and perhaps the entire Andean Ridge. This defines war as well as insurgency.
- The primary threat, then, is not an enemy military force (the PLA), although that is a serious challenge. The threat is not the debilitating political-economic-social instability generated by an irregular asymmetric aggressor, either. Rather, at base, the main threat is the inability or unwillingness of the government in office to take responsibility for and conduct legitimate measures to exercise effective sovereignty and to provide security and well-being for all its citizens. That governmental failure to protect and provide for the well-being of the people is what gives an aggressor the opening and justification for its existence and action.

- As a corollary, the ultimate threat is state failure or the violent imposition of radical political-economic-social restructuring of the state and its governance in accordance with the values—good, bad, or nonexistent—of the victor.
- Targeted regimes and their international allies that fail to understand the political-psychological ramifications of the new long-term guerrilla war and respond only with military and police power to the rhetoric, the phantom people's militias, and the myriad irregular methods to influence public opinion are not likely to be successful.

These points are all too relevant to the "new" political-psychological guerrilla wars of the twenty-first century. The lesson is straightforward: on so vast a field of action in which everyone and everything becomes involved, one is not grappling with conventional guerrilla military forces. Rather, one must contend with peripheral (indirect) war in which a few armed elements are acting clandestinely within a population and being directed by a specialized organization designed to create and control public opinion and action. In these terms, war will not disappear; it only changes. War thus becomes a juxtaposition of a multitude of small, carefully coordinated, political-psychological-economic-social-military actions. Intelligence and ruse, combined with propaganda and physical coercion, are the "new" instruments of state and nonstate power.[72] Qiao and Wang further remind us that this kinder, more gentle war in which bloodshed and brutality may be reduced is still war. "It may alter the cruel process of war, but there is no way to change the essence of war, which is one of compulsion, and therefore it cannot alter its cruel outcome either."[73]

CHAPTER 3

Four Trojan Horses of Different Colors

Vignettes from Al Qaeda in Spain, the Cuban Popular Militias, Haiti, and Brazil

Jorge Verstrynge makes a Leninist argument to the effect that internal war (that is, a nonstate actor versus a state within the nation-state) reflects a high level of social disequilibrium in the nation-state. That social disequilibrium can be exacerbated by internal and external forces.[1] One example, both internal and external, is that of immigration and migration. This is the situation regarding the nonterritorial Islamic communities in Spain and much of the rest of Europe. Until recently, the social violence connected with the Islamic communities in Europe was the result of a relative privation in which the immigrant and/or migrant generally remained outside the national economy, politics, and society and culture of the host country. More recently, however, that social disequilibrium—especially when supported by an external or internal hegemonic state or nonstate actor—has provoked ethnic and political agitation and violence that has been described as "explosive" and a type of "Trojan horse" that represents a new form of asymmetric war.[2]

Communities that are segregated politically, economically, and culturally from the mainstream of the host country provide an ideal environment for revolutionary Islam (Al Qaeda) and criminal gang activities. They provide a place and the basis from which to generate resistance against internal exploitation; a convenient place from which to recruit activists for propaganda and agitation efforts in the host country and abroad; and the human and monetary resources from which to circulate "terrorists" or "militants" throughout the world. The Islamic nonterritorial community is, thus, a prototype Trojan horse from which to surprise and confront "Western exploiters and enemies."[3]

Accordingly, a nonterritorial community can become a virtual state within the sovereign boundaries of a traditional nation-state. In such a virtual state, there is no need for a centralized bureaucracy, official armed forces, or a geographic heartland, only a small, flexible, and mobile group of cell-like fighting forces (gangs) able to conduct high-damage, low-cost actions calculated for maximum political-psychological-economic damage. There could also be a larger number of small low-profile support centers carefully located in a targeted society and capable of providing logistical, financial, manpower, sanctuary, and/or political support—on call. This agitation and propaganda (agi-prop) effort may be managed by an external or internal franchise-type organization providing guidance, expertise, and logistical and informational support as necessary.[4]

Additionally, there are other Trojan horses of very different types (or colors), including but not confined to the following two general categories: independent nonstate actors (for example, Haiti's gangs) and interrelated state or state-supported actors (for example, Cuban popular militias). The common denominators in these cases (vignettes) are as follows. First, any of these Trojan horses can begin to govern ungoverned terrain within a traditional nation-state and begin to acquire political power.[5] As a result, they tend to evolve into virtual states within a state. Second, these actors (whatever their type) "fuel a bazaar of violence where warlords, [drug barons], and martial entrepreneurs fuel the [ambiguous] convergence of crime, [insurgency], war, and politics."[6] Third, these actors work toward very similar, if not identical, strategic objectives. The putative objective in all these cases is to coerce radical political change in the physical or human terrain they seek to control.[7]

This is part of the larger Al Qaeda approach to asymmetric war (defensive jihad) in Western Europe. In that connection, we will look briefly at Al Qaeda's 2004 attack on the Atocha train station in Madrid, Spain. This action demonstrates the effectiveness of small cells and support centers in the role of Trojan horses and in creating psychologically devastating and politically transforming "divine surprises" in host countries. We will also consider the Cuban popular militias, which are intended to provide a second echelon of defense, after the regular armed forces, against an external invader or an internal actor that aims to take antisociety advantage of any political chaos that might come about as a result of the death or incapacitation of either or both of the Castro brothers. What is intended, however, may or may not come to fruition in a chaotic political-social environment. In any case, the Cuban paramilitaries should not be ignored. Then, we will assess the ubiquitous independent nonstate actor gangs in post–2009 earthquake Haiti. These gangs *(cacos)*, if left uncontrolled, are likely to contribute significantly to national and regional instability. As they evolve, they are capable of generating more and more terror, violence, and political-economic-social disequilibrium over wider and wider sections of the political map. Last, we will examine the devastating May 2006 "surprise" in São Paulo, Brazil, that was organized, controlled, and terminated by the leader of the Primeiro Comando da Capital (PCC) gang from a Trojan horse maximum security prison, using a mobile telephone. This vignette exemplifies a real and significant threat to the security, stability, effective sovereignty, and governance of Brazil—or any other relatively weak government.

These vignettes are intended to help civilian and military leaders—and the concerned public—to understand the potential power and capability of the gang phenomenon to generate widespread political-social disequilibrium. At the same time, these vignettes demonstrate the power of "phantom groups" all around the world to influence public opinion and transform domestic and foreign and defense policy. Strategic leaders must think about and deal with these destabilizing problems from multiple angles, at multiple levels, and in varying degrees of complexity. The alternative is to watch the global security arena become further engulfed in a chaos of violence, vice, corruption, and a lack of legitimacy.

AL QAEDA IN SPAIN: THE MADRID BOMBING, MARCH 2004

Before and shortly after March 11, 2004, Al Qaeda's asymmetric global challenge appeared to many to be ad hoc and senseless. Nevertheless, a closer look at the ruthless terroristic violence in Spain in March 2004 reveals some interesting and important lessons. On March 11, 2004, ten rucksacks packed with explosives were detonated in four commuter trains at Madrid's Atocha train station. That terrorist act killed 191 innocent and unsuspecting people and seriously injured over 1,800 more. That act was considered to be the most violent in Western Europe since the 1988 bombing of Pan American Flight 103 over Lockerbie, Scotland, that killed 270 people. Despite its length, the 1,470-page official summary of the investigation of the Madrid bombings provided very little information. It indicated that 29 men were involved in that attack: 15 Moroccans, 9 Spaniards, 2 Syrians (1 with Spanish citizenship), 1 Algerian, 1 Egyptian, and 1 Lebanese. The summary also indicated that the accused individuals were members of a radical political group active in North Africa and that Al Qaeda exercised only an inspirational influence. Moreover, the official summary indicated that these terrorists might have learned their bomb-making skills not from Al Qaeda but from the Internet.[8]

Subsequent British and other Western European investigations of terrorist attacks in Western Europe provided considerable additional information regarding the March 2004 bombings in Madrid and the twenty-nine-man organization that was responsible for that act. Those investigations indicated more than a casual relationship with Al Qaeda. Four of the bombers were Al Qaeda veterans from the base organization that provided leadership and expertise for the operation. Most of the nonveterans involved in the planning and implementation of the attack were operating as part of a lower, third ring of the base organization and were involved in criminal gang activities such as drugs-for-weapons exchanges, providing false documentation (passports, other personal identification, and credit card fraud), and jewel and precious metals theft.

Additionally, the nonveteran members of the gang were involved in disseminating propaganda and recruiting Spanish Muslim fighters to join Iraqi and other Al Qaeda–sponsored insurgencies. The intent of these day-to-day activities was to help support and fund regional and global Al Qaeda jihadist operations.[9] In this instance, the normal criminal gang activities of the twenty-nine-man group were interrupted so that they could take on the mission of bombing the Madrid train station.[10] Not until the bombing of the Atocha station, then, did this particular gang transition from an implicit political agenda (that is, recruiting personnel and criminally generating financial support for Al Qaeda's political-military operations in the Middle East, North Africa, and elsewhere) to an explicit political challenge to the Spanish state and the global community. Therefore, that was the point at which these "delinquents" became "militants." The purpose of the action was not to achieve any military objective, and it was not a random act. Instead, the bombing was deliberately intended to generate strategic-level political-psychological results. (Nevertheless, the militancy continued to be treated as a social and law enforcement issue.)[11]

What This Propaganda-Agitation Effort Demonstrated

Since March 2004, Al Qaeda has demonstrated that it can skillfully apply irregular asymmetric war techniques to modern political war and has done so with impunity. Indeed, its terroristic actions were executed in a way that made virtually any kind of Spanish, Western, or U.S. military response impossible. After over three years of investigation and the trial, the Spanish court acquitted seven of the twenty-nine defendants and found twenty-one individuals guilty of involvement in the 2004 Madrid train bombings. (One of the accused had been previously convicted on charges of illegal transporting of explosives. Also, four of the twenty-nine accused men committed suicide three days after the bombing.) Two Moroccans and a Spaniard were sentenced to 42,924 years in prison. Nobody else in the gang was sentenced to more than 23 years in prison. Of great importance is the fact that the men accused of planning and carrying out the attack were not convicted for the train bombing:

they were found guilty of belonging to a terrorist group or for illegally transporting explosives.[12]

The Madrid attack also sent several messages to the Spanish people, the rest of Europe, the United States, and Muslim communities around the globe. The various messages went something like this:

> Supporting the United States in its global war against terrorism and in Iraq is going to be very costly.
> Countries not cooperating with Al Qaeda might expect to be future targets.
> Much can be done with a minimum of manpower and expense.
> Al Qaeda is capable of moving into the offensive against European and other Western enemies, should the organization make the political decision to do so.
> Bombings can be deliberately executed in a way that makes any kind of Western or U.S. military response impossible.
> Al Qaeda can successfully stand up against the United States and its allies.[13]

As a result, the publicity disseminated throughout the Muslim world has been credited with generating new sources of funding, new places for training and sanctuary, new recruits to the Al Qaeda ranks, and additional de facto legitimacy.[14]

Additional Strategic-Level Results of the Bombing

Even though the information gathered throughout Western Europe from the investigations and trials connected with the Madrid bombing was treated cautiously and without alarm, the results achieved by the twenty-nine-man cadre (gang) were dramatic and significant. The sheer magnitude and shock of the attack changed Spanish public opinion and the outcome of the parliamentary elections that were held just three days later. In those elections, the relatively conservative, pro-U.S. government of Prime Minister Jose Maria Aznar was unexpectedly and decisively defeated. That defeat came at the hands of the anti-U.S./anti–Iraq War leader of the socialists, Jose Luis Rodriguez Zapatero.

Prior to those elections, the Spanish government had been a strong supporter of the United States, U.S. policy regarding the global war on terror, and the Iraq War. Shortly after the elections, Spain's 1,300 troops were withdrawn from Iraq, and consensus at top levels of the parliament had it that Spain would no longer be a strong U.S. ally within the global political and security arenas.[15]

Long-Term Objectives

These political-psychological consequences advanced the intermediate and long-term objectives of political war that Osama bin Laden and Al Qaeda have set forth. The most relevant of those objectives, in this context, are intended to erode popular support for the war on terrorism among the populations of American allies and to gradually isolate the United States from its allies. All that—along with the messages noted above—was accomplished by a small, twenty-nine-man gang, at a cost of only $80,000.[16]

CUBAN PARAMILITARY ORGANIZATIONS (PEOPLE'S MILITIAS) IN THE POST-CASTRO ERA

Armed nonstate groups all over the world are directly challenging targeted governments' physical and moral right and ability to govern. This almost chronic chaos is exacerbated by nation-state actors and nonstate actors using nonstate popular militias, youth leagues, gangs, or their equivalents to help take control, maintain control, or regain control of a given political-economic-social entity. It is in this context that popular militias have been organized, trained, and nurtured in Cuba. For fifty years, Cuba's popular militias (paramilitary organizations) have been expected to act as "midwi[v]es for new social orders" (as they did in Africa during the 1960s and early 1970s) and to help defend and maintain the revolutionary Cuban socialist state.[17] Anyone contemplating the post-Castro era in Cuba will certainly have to take these paramilitary organizations into consideration.

Background: The Cuban Armed Forces and Paramilitary Forces (Popular Militias)

The regular (conventional) Cuban armed forces (Fuerzas Armadas Revolucionarias, or FAR) have recently been reduced in size and now consist of 49,000 active-duty personnel and over one million reservists. These forces are conventionally divided into the army, navy, and air force and are under the political control of the minister of the Revolutionary Armed Forces (MINFAR). The paramilitary forces consist of 26,500 personnel and are divided between the Youth Labor Army (Ejercito Juvinil de Trabajadores, or EJT), the Territorial Militias (Milicias de Tropas Territoriales, or MTTs), and the Committees for the Defense of the Revolution (Commites de Defensa de la Revolucion, or CDR). These paramilitary organizations fall under the political control of the minister of the interior (MININT). The intent of such a division of labor is to create multiple security organizations that are ultimately responsible to the supreme leader of the state through the MINFAR and the MININT.[18] As such, and under their respective immediate political masters (depending on who these might be in a socialist succession or a democratic transition), they theoretically act as counterfoils to each other. At the same time, this division of labor is intended to ensure that the supreme leader of the Cuban state cannot be challenged by a single, all-powerful military entity. Thus, legally, and organizationally, the leader (whoever he may be) maintains and controls a complete monopoly of state power.[19]

The FAR is structured, trained, and located to accomplish two missions: the conventional defense of Cuba against U.S. aggression and providing internal security. The paramilitary forces (EJT, MTT, and CDR) are organized and trained to provide strategic depth from which to regain control of the country if attacked by an external aggressor. These paramilitary organizations are also prepared to fight a *guerra de todo del pueblo* (guerrilla or peoples' war) in the event of conventional defeat by an external aggressor or antirevolutionary activities generated by an internal actor. Thus, doctrinally, even if the FAR should be defeated by an internal or external enemy force, the popular militias would be in a position to continue to fight to defend or reestablish the threatened socialist state.[20]

Accordingly, training, doctrine, and inculcation of a revolutionary culture in the Cuban paramilitaries preclude the concept of a new and different (for example, liberal democracy) type of government, as well as the dissolution of the current security apparatus. That is, either there will be no change in succession to another socialist regime or there will be a period of political turmoil in transition to a new and different type of government. Theoretically, that chaos would be expected to be similar to that experienced in Russia after the Revolution of 1917 and before the consolidation of power by Lenin. It also reminds us that the contemporary Russian Nashi organization is based on the same Leninist concept (see chapter 4). This takes us to a further elaboration of the doctrinal role of the paramilitaries in Cuba after the end of the Castro era.[21]

Elaborating the Role of the Cuban Paramilitaries in the Event of a Post-Castro Period of Turmoil

Left in dire straits, subject to depredation, and denied adequate and legitimate access to basic services and personal security, people become susceptible to the exhortations of demagogues, hatemongers, revolutionary reformers, transnational criminals, and other armed individuals who might want to take control of the state. The concomitant political-economic-social-informational-military turmoil can allow peoples' militias (paramilitary organizations) to operate with relative freedom of movement and action. In these terms, conflict and adversaries will have changed. Conflict will no longer be only an instrument of state action but will also consist of nonstate engagement. As a result, the adversary (and center of gravity) will no longer be solely a recognizable military force and a nation-state's industrial-technical capability to support military operations. Rather, new adversaries will include nonstate actors and their various allies who can influence and control public opinion and political decision-making leadership. The basic reality of this new center of gravity is that information and the media—not military firepower or technology—serve as the primary currency upon which a contemporary people's war is conducted.[22]

In this type of security environment, the doctrinal purpose of popular militias is to raise the level of direct popular action against "indigenous feudalism, aboriginal capitalism, sepoyan militarism, and *yanqui* imperialism." The intent is not to destroy an enemy military force. The objective of the paramilitary effort is—with popular support—to wear down the enemy government and people to the point where their resolve is dead. There is no need for conventional military forces. There are no maneuver forces, no design for conventional battle, and no obvious connectivity with other paramilitary actions going on elsewhere. Each engagement is particular to itself but theoretically connected through a system of networks and an overarching political ideology—to depose an enemy regime and provide the leadership to reestablish a Cuban socialist state. Thus, "Political and moral factors are more decisive for victory than heavy armament and ironclad units."[23] Doctrinally, the Cuban paramilitary organizations will accomplish their objectives over time and in a minimum of seven phases:

1. Organize and train cadres of professionals for political leadership duties, political-psychological-military combat, and the creation of selected environments of chaos;
2. Create a popular political party out of individuals of all persuasions who will work together to disestablish opposed societies and establish a new social democracy;
3. Plan and execute covert and overt persuasion, intimidation, and coercion activities against targeted individuals and institutions;
4. Organize, train, and develop local popular militias to fight and to defend their own localities;
5. Foment regional conflicts;
6. Confront—gradually—demoralized external or internal enemy military forces and act as a catalyst to bring about their desired collapse; and
7. Reestablish socialist governance.

This is not the rhetoric of a disappointed old man who will never see his revolutionary dreams fulfilled. Rather, this is the rhetoric of an individual performing the traditional and universal Leninist function of providing a strategic vision and an operational plan for maintaining or regaining revolutionary power.[24]

The Cuban Security Environment in the Long Term

The security environment in the post-Castro era can either remain quite stable, in totalitarian terms, or become extremely volatile and dangerous. The precise outcome would depend to a large extent on the level of chaos and the number and types of parties involved in that turmoil. Additionally, unlike their Nashi cousins in Russia, the Cuban popular militias have not been purposefully tested over time since their involvement in Africa over forty years ago. Thus, whether the contemporary Cuban popular militias can or will perform their assigned missions is not known. What is known from the lessons of history, however, is that in the event of invasion from abroad, serious internal instability, or a coup, popular militias are likely to devolve into small, flexible, mobile, and undirected cell-like fighting forces (gangs). Moreover, they will probably look to secure their own interests, rather than those of a defunct regime—and further exacerbate the internal instability situation. Of course, if another external actor (such as Venezuela) or an internal actor (such as a junta) could somehow impose direction and discipline on those militias, they could become what they are intended to be—formidable counterrevolutionary forces. In any event, the Cuban paramilitaries will warrant careful consideration.

THE CHALLENGE OF HAITI'S FUTURE

An understanding of the historical-political background within which Haiti operates must begin with three controlling realities: Haitian history and politics, Haiti's cultural values, and U.S. interventions. First is the political reality of Haiti's 190 years of independence. That period from 1804 to 1994 has been characterized by chronic poverty, political instability, violence, and foreign intervention. Second, the social reality of Haiti's caste system and its values cannot be ignored. Third, the political-historical reality of three interventions by United States (1915–34, 1994–2000, and 2004) in the country is instructive. These contextual realities begin to explain the challenge of Haiti's future and the fact that Haiti has been a predatory state for more than two centuries. The implications for the political, police, and military forces that must face that kind of criminal anarchy are volatile and dangerous.

Historical-Political Context: The First 190 Years, 1804–1994

After a dozen years of bloody revolutionary war against the French, free Haiti emerged in 1804 in control of the western side of the island of Hispaniola. Napoleon Bonaparte's expeditionary force of 43,000 men had been defeated in 1803 by an ill-equipped army of illiterate ex-slaves under the command of Toussaint L'Overture, a slave grandson of an African king. He was captured, however, and sent to prison in France. A year later, in 1804, self-appointed general Jean Jacques Dessalines defeated the last of the French forces and declared the independence of Haiti. He crowned himself as Emperor Jacques I and put the now theoretically free peasants to work at tasks as onerous as they had ever known as slaves. After two years, the populace turned against the emperor-for-life and he was shot. Next came the last of the illiterate revolutionary generals, Henri Christophe. Like Dessalines, he had himself crowned, as King Henri I, and compelled the peasants to work under armed guard. He lasted in office until 1808. His successors emulated his ineffective rule until 1843, when President Jean Pierre Boyer was forced into exile.[25]

Between 1843 and 1915, twenty-two dictators rose and fell, virtually continuous civil violence shook the country, and the masses sank to new depths of misery. As might be expected from unenlightened dictatorial rule, and despite a few building projects, it was a period of economic and social stagnation. Over 90 percent illiteracy existed throughout Haiti, and there was no sense of civic responsibility, no institution building, no rule of law (a pledge of allegiance to Dessalines, as an example of the standing law, went something like this: "I swear entire obedience to the laws he shall deem fit to make, his authority being the only one I acknowledge, I authorize him to make peace and war, and to appoint his successor"). The only laws that seemed to be universally accepted were those ending slavery and those prohibiting a white person from owning property in Haiti. In that connection, the large landholdings that had belonged to the French colonists were broken up and distributed in very small (and unproductive) plots to everybody.²⁶

Social-Cultural Context: The Caste System in Haiti

In the first 190 years of Haitian independence, a cultural issue in the form of a caste system developed and took hold in the society. Early in Haiti's independence period, Dessalines ordered the execution of the few whites who were left from the revolution. There were about 30,000 *gens de couleur,* the mulattos who had been freed under French rule. There were also a half-million blacks (*negros*). Thus, the two classes—the elites and the masses—that emerged from the revolution continue to survive. The elites, never more than 2 percent of the population, are the descendants of the *gens de couleur*—the freedmen of the colonial days who took control of the landholdings and wealth of the French colonists. The blacks were and are the vast majority of the population, and after 1843 black leaders took political control of the country. The elites were forced to comfort themselves with the wealth they had accumulated and with a sense of social superiority. Consequently, there remains a deep gulf between the elites and the masses. The masses live much as their slave progenitors lived in the eighteenth century: they are almost entirely illiterate; their language is Creole, not French; Voodoo *(Vodun)* is their religion, not Catholicism; and their vocation is virtually anything that does not require literacy, training, mobility, or capital. In that political-social-economic milieu, Haiti has built no strong political institutions but instead has existed in a state of permanent political-social-economic chaos.²⁷

Historical-Political Context: The Interventions

Prelude to the U.S. Intervention in 1915. Clearly, Haitian politics had always been violent. Few presidents left office alive. It became easier and easier for aspirants to national power to recruit a force of unemployed peasants from the nation's north, provide them with rudimentary arms, and encourage them to assail the capitol city of Port-au-Prince, looting as they went. Known as *cacos,* most would then return to the north, where they were available for the next would-be president. In July 1915, another *caco* army descended on Port-au-Prince. Before fleeing to the French Legation, President Guillaume Sam had 161 political prisoners, many from elite families, murdered. An enraged mob broke into the legation and literally tore Sam apart.[28]

The U.S. Intervention of 1915–34. As a result of the country's chronic political-economic-social turmoil between 1843 and 1915, seven men, including Guillaume Sam, had violently seized control of the Haitian presidency. That instability, along with the related civil unrest, the fiscal irresponsibility generated by the various governments, and the possibility of a German naval base in Haiti, brought U.S. president Woodrow Wilson directly into Haitian politics. That, in turn, brought U.S. Marines into the country—and they stayed for nineteen years. During that period of "occupation," U.S. officials directed or controlled virtually everything the Haitian government did. Important projects that had been ignored or neglected for over one hundred years were begun—and some were completed. Among other things, the omnipresent graft was eliminated, public health programs were promulgated, and an army and police force were created. But Haitians (except for puppets) were left out of the decision-making and policy-making processes, and Haitians, understandably, found reasons and means to resist virtually every U.S.-sponsored measure. As a consequence, the U.S. intervention and tutelage did not bring the stability, development, well-being, and democracy that Americans had worked for and expected.[29]

The U.S. Intervention of 1994. In 1994, when the United States invaded Haiti to reinstall coup-ousted but democratically elected President Jean-Bertrand Aristide to power, American forces faced militia-style gangs (cacos) loyal to the coup leader, General Raoul Cedras. The decision to restore Aristide to the presidency and the implementation of this decision were heavily influenced by the U.S. focus on democratic procedures and the War on Terror. The War on Terror in Iraq and Afghanistan, however, severely limited what the United States was capable of doing in Haiti. Thus, the United States was unwilling to undertake a long-term effort to establish respect for democracy, human rights, and honest administration under the United Nations (UN) mandate. As a result, the decision was made to intervene with a small force to achieve political stability in the short term, and enlist the support of countries within the Western Hemisphere and the Organization of American States (OAS) to deal with Haiti's long-term future.[30]

A small number of U.S. forces remained in Haiti for six years, until the flawed parliamentary elections and meaningless presidential elections of 2000. Then, after condemning the 2000 elections, the OAS left the country. The UN pulled its mission out of Haiti in 2001—complaining that that country had no governmental institutions with which to work. The hope had been that Haiti would raise itself from being the poorest country in the hemisphere to emerge as a modern democratic society. Rather, the country experienced increasing demagoguery, lawlessness, gang *(caco)* activity, impunity from punishment for lawlessness, and rampant illegal drug trafficking.[31] Clearly, hope is not a strategy.

U.S. Intervention in 2004. Between 2001 and 2004, conditions in Haiti worsened. The downward spiral of graft, poverty, lawlessness, violence, instability, impunity, and drug trafficking continued. By early 2004, the whole of Haiti was being terrorized and controlled by armed thugs and gangs. With the support of gangs, opposition groups were able to defeat the Haitian National Police, capture the key towns around Port-au-Prince, and begin to pressure still-president Aristide to resign. As the opposition cacos moved into Port-au-Prince, pro-Aristide gangs set up blockades and then began looting and disappearing into the slums of the city. Aristide finally understood he was defeated and wrote a letter of resignation. He also requested U.S. assistance to depart Haiti safely. The president of the Supreme Court assumed the office of president of Haiti and requested the UN to provide a military force to stabilize the country. Accordingly, the Security Council passed a resolution authorizing a multinational stabilization force for ninety days that, in turn, would prepare for a follow-on UN force. Almost immediately, a U.S. Marine Air and Ground Task Force (MAGTF) arrived in Port-au-Prince—followed by French, Chilean, Canadian, and Brazilian troops. A UN stabilization force (MINUSTAH), led by Brazil, has been in Haiti since that time.[32]

Where Haiti's History Leads

This takes us to the disastrous earthquake of 2009. Amid the chaos of the earthquake, it might seem premature to think about indigenous forces the United States and its international partners might have to confront in Haiti during the "rebuilding" period. Actually, it is vital. Time is short before the world's generosity turns to cynicism and corrupt Haitian public officials, civilian contractors, and caco leaders turn their high expectations for self-enrichment to violence. Time is short before the suffering of the past two centuries—to say nothing of the earthquake—risks being repeated.

We are already aware of the presence of gangs and their involvement in the politics of Haiti. As one example, gangs were fundamental to Aristide's success at staying in the Presidential Palace for ten years. At the same time, opposition cacos were fundamental to bringing Aristide down. Clearly, powerful individuals and elites from across Haiti's political spectrum exploit gangs as an instrument of political warfare, providing them with arms, funding, and immunity from the law. Now, another dynamic must be added to the Haitian political equation. After Aristide's downfall, gangs that supported his political opposition assumed more independence and stepped up their involvement in the big business of illegal drug trafficking. Additionally, many towns and villages and the rural areas of Haiti have had the benefit of little or no consistent government or UN or U.S. presence. This has left the control of vast areas of the country, including the larger cities, to armed groups that, more likely than not, are developing close ties with transnational criminal organizations (TCOs) heavily involved in the very lucrative narcotics transshipment business.[33]

The equation that links gangs, TCOs, and insurgents turns on a combination of need, organizational infrastructure, development, ability, and the availability of sophisticated communications and weaponry. For example, traffickers possess cash and lines of transportation and communications. Gangs possess followers, discipline, and organization. Traffickers need these to help protect the assets and project their power within and among nation-states. Gangs (cacos) are in constant need of logistical and communications support, arms—and money.

Both groups possess relatively flat organizational structures that, when combined, can generate more efficient and effective decision making than any slow-moving bureaucratic, hierarchical governmental system. This is an important element of power in itself and gives the gang-TCO phenomenon the capability to seriously challenge the Haitian state. In this connection, the primary objective of gangs, TCOs, and insurgents is to attain a level of freedom of movement and action that allows the achievement of their desired enrichment and control of government. This defines insurgency: that is, to neutralize, control, or depose a government. Rephrased slightly, it also defines war: that is, compelling an adversary to accede to one's primary objective.[34]

This nexus represents a triple threat to the authority and sovereignty of a government. First, murder, kidnapping, intimidation, corruption, and impunity undermine the ability and the will of a state or nonstate opponent to perform its legitimizing security and public service functions. Second, by violently imposing their power over bureaucrats and elected officials of a state, the TCOs and elements of the gang phenomenon compromise the exercise of authority and replace it with their own. Third, by neutralizing (making irrelevant) governance and taking control of portions of the national territory, and perhaps performing some of the tasks of government, the gang-TCO phenomenon can de facto transform itself into quasi-states within a state. In addition, the criminal leaders enjoy freedom of movement and govern those areas as they wish.[35]

The resultant convoluted array of gangs and TCOs (that is, the gang-TCO phenomenon) leaves an anarchical situation throughout Haiti. As each caco and TCO violently competes within itself and with others and works against the government and its international supporters to maximize its share of freedom of movement and action and profit, we see a strategic internal security environment characterized by ambiguity, complexity, and unconventional (irregular) war. We also see the further erosion of the Haitian state and the establishment of small and large "zones of impunity" throughout the country. This situation reminds one of the feudal medieval era. Violence and the fruits of violence—arbitrary and unprincipled political control—seem to be devolving to small, private, criminal, feudal, and Trojan horse–type nonstate actors.

This is one more serious challenge to democracy, security, stability, well-being, and sovereignty in Haiti.[36]

Implications for Governments and Their Political, Military, and Police Forces Facing Criminal Anarchy

Despite official optimism regarding Haiti's future, the internal security environment that is developing in Haiti today goes well beyond a simple law enforcement or force-on-force military problem. What we see is an unconventional battlefield that no one from the traditional-legal Westphalian school of conflict would feel comfortable dealing with. Instead of conventional war conducted by uniformed military forces of another country, we see something more political-psychological, complex, and ambiguous. We also see something that cannot be fixed with the same kind of high-tech power that enabled the United States to take control of Afghanistan and Iraq in a very short time and with negligible losses. The old paradigm was that of interstate industrial war. The new one involves war "amongst peoples." War among peoples reflects some hard facts:

- Combatants tend to be small groups of armed soldiers who are not necessarily uniformed, not necessarily all male but also female, and not necessarily all adults but also children;
- These small groups of combatants tend to be interspersed among ordinary people and have no permanent locations and no identity to differentiate them from the rest of a given civil population;
- Armed engagements may take place anywhere;
- Combat or confrontation uses not only coercive military-type force but also co-optive political, psychological, and economic persuasion;
- The major military and nonmilitary battles in modern conflict take place among the people, and when they are reported, they become media events that may or may not reflect social reality; and
- All that is done is intended to capture the imaginations of people and weaken the will of their leaders.

Additional strategic-level analytical commonalities in the contemporary battle space would include the following:

There are no formal declarations or terminations of war;
There is no specific geographical territory to attack or defend;
There is no single credible political actor with which to deal, and no guarantee that any agreement between or among contending actors will be honored;
There are no national or international laws, conventions, treaties, or boundaries that cannot be ignored or utilized; and
There are no instruments of power that can be ignored or left unused.[37]

Restructuring the Haitian State

Given the past experience of Haiti, an analysis of the current strategic security environment indicates that the ultimate threat of destabilizing caco activities is not political-social disequilibrium or criminal violence. It is not even state failure or the coerced imposition of a radical restructuring of the state and its governance. Sooner rather than later, the nations and international organizations engaged in rebuilding Haiti will be forced to address the values that determine the *quality* of governance, security, and stability. One set of values serves cruel criminal greed. The other seeks to serve the general well-being.

THE BRAZILIAN PRIMEIRO COMANDO DA CAPITAL (PCC)

The great city of São Paulo, Brazil—the proverbial industrial "locomotive" that pulls the "train" of the world's eighth largest economy—was paralyzed by a great, if not divine, surprise in mid-May 2006. Virtually nothing moved for five days. More than 293 attacks on individuals and groups of individuals were reported, hundreds of people were killed and wounded, and millions of dollars of damage was done to private and public property. Busses were torched, banks were robbed, personal residences were targets of violence, municipal buildings and police stations were attacked, and rebellions broke out in eighty-two prisons within the State of São Paulo's penal system. Transportation, businesses, factories, offices, banks, schools, and shopping centers were shut down. In all, the city of São Paulo was a frightening place during those five days in May.[38]

During that time, the PCC demonstrated the ability to coordinate simultaneous prison riots; destabilize a major city; manipulate judicial, political, and security systems; and shut down the formal Brazilian economy. The PCC also demonstrated its lack of principles through its willingness to indiscriminately kill innocent people, destroy public and private property, and suspend the quality-of-life benefits of a major economy for millions of people. Beyond security forces—which were reportedly as involved in extrajudicial killings as were the criminal perpetrators of the chaos—the violence and chaotic conditions in São Paulo made any effort to assert governmental authority or conduct essential public services virtually impossible.[39]

PCC Organization and Motives

The PCC has an estimated 65,000 to 125,000 full- and part-time dues-paying members and is led by a brilliant and uncompromising career criminal called Marcola (Marcos Williams Herbas Camacho). Although analysts believe that not more than 6,000 active PCC members are in Brazil's prison system, they know that the PCC has extended its influence into the favelas, or "ungoverned slums," in São Paulo, Rio de Janeiro, and other major cities of Brazil. This has been accomplished through a long series of carefully negotiated, sometimes forced alliances with other gangs and favela chiefs (*chefes da favela*). As a result, at any given time, Marcola controls at least 60,000 PCC members in the prisons and favelas of the country.[40]

Ostensibly, this turmoil and retribution was triggered by prisoners who were being transferred to a maximum security prison that was not equipped to allow the inmates to watch the much-anticipated World Cup soccer matches on TV. Thus, an ambitious, prisoner-initiated "prison rights" agenda was the motive for the rebellion. But at its base, consensus has it that the surprise May explosion in São Paulo was really a show of force by the largest criminal gang in the Western Hemisphere. The primary intent was to announce to the state and federal governments that the PCC and its allies in the favelas were strong enough to compel the negotiations of terms of state sovereignty vis-à-vis that organization. Unlike many gangs in other parts of the world that seek to permeate government to the point where the state authorities and selected gang members are the same people, the PCC has attempted to neutralize the Brazilian state within its sphere of influence. At the least, given that Marcola got everything he wanted out of the negotiations to end the chaos in São Paulo, it is probably safe to say that the PCC and the *chefes* or barons of the favelas have grown more powerful, and the state relatively more constrained.[41]

The Program of Action

Favelas are the bases of the PCC's extended power. In the favela, "traffic" is everything, and "territory controlled" is critical. The PCC, like other criminal gangs throughout the Western Hemisphere and the world, is deeply involved in drug trafficking, arms trafficking, human trafficking, murder, kidnapping, robberies, and extortion. To maintain its momentum and expand its markets, the organization has increasingly adopted an offensive mode with tactics appropriate to urban guerrilla war, in which it looks for confrontations with rival gangs and police and military forces. PCC members and temporary-hire "soldiers" from the favelas carry out their violent tasks armed with automatic weapons, machine guns, hand grenades, rocket-propelled grenades, antipersonnel mines, and crudely armored vehicles. Command and control is provided primarily through a very efficient communication network based on mobile telephones. (This takes us back to Marcola and his cell phone.) In areas controlled by the PCC or in areas that might be "invaded" by PCC-controlled units, one has a choice: to pay dues, mentally submit, and physically contribute to the organization, or *"subir al cielo"* (to die).[42]

In addition to its violent turf-controlling efforts and illicit trafficking activities, the PCC pursues more than a casual, self-serving criminal rights agenda. The organization hires eighteen to twenty lawyers who work full-time. They act not only as advocates for gang members but also as mentors for young gang members. One of the great successes of the PCC has been to infiltrate, or "colonize," the governmental organizations that administer the entrance examinations necessary to enter the Brazilian public service. The job of the PCC lawyer-mentor is to ensure that young gang members (and children of the convicts) who have the ability and the desire to enter into public service can and do get the necessary education and pass the appropriate examinations. As a consequence, the PCC is putting its own people into bureaucratic positions it considers important in the Brazilian system. Thus, in addition to controlling slums in the major cities of the country, the third-generation parts of the PCC appear to be slowly and surely extending their influence into the public service. The logical conclusion regarding this effort would be, simply, that Marcola is deliberately leading his organization to infiltrate the state. This, of course, would be an important objective in the process of securing freedom of movement and action and in moving Brazil toward criminal-state status.[43]

Response

The São Paulo state government and the Brazilian federal government appear to have been relatively unconcerned with the specific issues that brought on the May 2006 crisis. The official State of São Paulo response to the violence and chaos was simply, "I say to our people that the police are still in the streets, [the people] can go out and have fun this weekend."[44] This "business (or fun) as usual" approach to the gang problem is similar to that expressed not too long ago when a high-ranking federal official said, "Not to worry. Brazil will grow out of this."[45]

On the positive side of this dilemma, the unfortunate São Paulo 2006 "surprise" brought to light socioeconomic-political-psychological problems—poverty, corruption, penetration of the political system, and impunity—that should be debated sooner rather than later and result in something more than simply "tough talk." In that connection, the Brazilian people have demonstrated their displeasure with the "business as usual" (official lassitude, inefficiency, and outright corruption) approach to dealing with the PCC and other gangs. As an example, reportedly, citizens of Rio de Janeiro (called Cariocas) "rejoiced" as the usual hectic pace of murder, assault, and theft slowed to almost negligible proportions when President Luiz Inacio Lula da Silva responded to public pressure and announced that 75 percent of the military and police equipment brought into Rio during the Pan American games in July 2007 would remain in the city. How that equipment will be used over time remains to be seen, but Cariocas have been reminded of what it feels like to live in a safe city.[46] On the negative side of this dilemma, the most serious point that must be made is that vigilante militias are violently beginning to impose their own "peace" in favelas the police do not control.[47]

Self-Enrichment and Impunity

The 2006 "surprise" organized by Marcola and the PCC from their Trojan horse maximum security prison in the State of São Paulo illustrates that loosely governed countries and ungoverned territories within them are attractive venues for gangs and other nonstate actors who seek to avoid the reach of criminal justice systems and evade the rule of law for their own advantage. Ironically, Marcola and his fellow PCC prisoners have found safe places for conducting their unprincipled self-enrichment activities. The May 2006 incident in São Paulo is a prime example of a new "urban jungle," within which gangs and their drug baron patrons and insurgent cousins can find political space from which to conduct their illicit, commercial enrichment operations.[48] This mixing of political and commercial interests is a lethal combination that exemplifies a real and significant threat to the security, stability, and effective sovereignty of the Brazilian state.

KEY POINTS AND LESSONS

What makes the above cases or situations and their implications significant beyond their own domestic political context is that they are situations from which lessons from contemporary asymmetric warfare can be learned. These cases are both the results and the harbingers of much of the ongoing political chaos of the twenty-first century. They stress the following:

- Through astute application of terrorist acts, groups (gangs) of irregular "militants" can have a disproportionate effect on public opinion and can compel fundamental changes in domestic, foreign, and defense policy.
- In a revolutionary or counterrevolutionary situation, there is no need for conventional armed forces. Small "phantom groups" of militia can act independently but still remain connected with others through a system of networks and an overarching political idea. The intent might be to depose an enemy regime and/or to reestablish the previous state. In Lenin's terms, popular militias (agi-prop gangs) can act as "midwi[v]es for new social orders."[49]
- Gangs—working from Trojan horse "zones of impunity," nonterritorial communities, or any other kind of virtual state within a state—represent a significant threat to the authority of a targeted government, and to those of its neighbors.
- Small groups operating as a part of the gang phenomenon, and from the sanctuary of some sort of Trojan horse, fuel the ambiguous convergence of crime, insurgency, war, and politics.
- Thus, the civil-military relations problem regarding the question of whether or not the gang phenomenon is a law enforcement issue or a national security issue is irrelevant. It is an issue larger than either. It requires the holistic application of all the instruments of power of the targeted nation-state and its international allies.

Sun Tzu reminds us that we do not need an abundance of manpower, specialized equipment, and financial resources to deal with an enemy such as the protean gang phenomenon. He recommends, more than anything else, the well-considered application of "brain power."[50] More recently, Qiao and Wang have pragmatically elaborated on that concept: "Warfare is no longer an exclusive imperial garden where professional soldiers alone can mingle. Non-professional warriors (hackers, financiers, media experts, software engineers, etc.) and nonstate organizations are posing a greater and greater threat to sovereign nations. Consequently, these new kinds of warriors must now be included in a nation-state's architecture for conducting war."[51]

CHAPTER 4

STATE-SUPPORTED INTERNAL AND EXTERNAL PERSUASION AND COERCION

The Russian Youth Group Nashi

Conflict (war) is no longer a simple state versus state, military-to-military confrontation. Conflict now purposely blurs the distinction between and among crime, terrorism, and warfare. It involves entire populations, as well as a large number of indigenous national civilian agencies, other national civilian organizations, international organizations, nongovernmental organizations, private voluntary organizations, and subnational internal actors involved in dealing politically, economically, socially, morally, criminally, and/or militarily with perceived ambiguous and complex threats to national security and well-being. In these terms, war is no longer only a violent confrontation between nation-states but has come to include a broad spectrum of violence within a given state that ranges from propaganda and agitation activities to civil violence to revolutionary uprisings (insurgencies) to civil wars to wars of national liberation. Many of these external and internal conflicts are initiated and maintained on behalf of causes espoused by governments of states wanting to maintain plausible deniability for any kind of aggression or to disclaim obvious responsibility for the consequent disequilibrium and erosion of a targeted society.

To this end, Nashi ("ours"), the most recent in the short list of Russian youth organizations that have been organized, reorganized, or reestablished since Lenin took control of the Soviet state in 1917, "burst into public view" in 2005.[1] The intent of Nashi was to build metaphorical Trojan horses (see chapter 3) modeled on the contemporary Cuban paramilitary organizations and to generate divine surprises, influence public opinion at home and abroad, and defend Russia from "a Western plot to encircle Russia."[2] Nashi and other state-supported groups like it represent an important type of irregular player in internal and external conflict. These actors are playing an increasingly sophisticated role in creating asymmetric indirect threats to stability, security, and effective national sovereignty all around the world today. Lessons derived from a brief examination of Nashi activities demonstrate how a state-supported set of "gangs" might fit into a holistic state or nonstate actor's effort to compel radical political-economic-social change, and explain how some traditional political-military objectives may be achieved indirectly rather than directly.

A first step in that direction is for strategic leaders to understand that an enemy cannot only be the military formations a state or nonstate actor may be able to put into the field. Rather, the enemy is now everyone and anyone who supports the "bourgeois class," or its equivalent, in any way. Thus, the enemy can be a nation-state, a nonstate actor, an internal or external institution, or even an individual who might conceivably be able to threaten a Leninist social-democratic order or its equivalent.[3] As a consequence, the conventional state-centric center of gravity (the source of all power on which everything depends) can no longer be a state or easily identified state-supported military or paramilitary formations. Instead, it must include external and internal decision-making leadership and public opinion.[4]

The basic reality of this alternative center of gravity, as demonstrated in the Cuban paramilitary model (chapter 3), is that information (propaganda) and agitation (persuasion and coercion) constitute the primary currency by which modern "war amongst the people" is run.[5] In these terms, one is not restricted to conventional direct military and paramilitary ways and means of conducting conflict. One must begin the process to compel political change and the adoption of new values (or restore old systems and values) using indirect persuasive-coercive ways and means, not just uniformed soldiers and recognized military formations.[6]

THE THEORETICAL-POLITICAL CONTEXT OF NASHI FUNCTIONS AND ACTIONS

Lenin's teachings provide the theoretical-operational basis of revolutionary Marxism. That, in turn, articulates the strategic asymmetric-irregular-political vision within which Nashi and many other contemporary state and nonstate actors now operate. It explains new and different definitions of "defensive war," "democracy," and "peace." It also explains the role of a state-supported set of "gangs," such as Nashi, to systematically compel radical change.

Defensive War, Peace, Democracy, and Revolution

The intent of a Marxist-Leninist revolution is to capture the will of a people and its leadership and bend that will to achieve radical political-economic-social change and implant new socialist values in a society. That political objective can be attained only as a result of what Lenin called a "defensive war." Such a war is conducted to protect the homeland and its external interests. A defensive war may also be conducted against internal enemies of the state.[7]

Lenin's primary assumption in that regard is that any social democracy (socialist state) is at considerable risk as long as even one bourgeois state (liberal democracy) or one bourgeois person continues to exist. He argued, "It can only be with the extinction of all [bourgeois states and individuals] that socialism comes about, as well as true sovereignty. Then you will be able to start building the edifice of communist society, and bring it to completion."[8] The destabilization and erosion of an enemy state is accomplished by the systematic (dictatorial) application of some combination of indirect and direct, state and nonstate, military and nonmilitary, and lethal and nonlethal actions.[9] Only a social democracy, or its clients, can conduct this kind of war. Only a social democracy can represent the democratic will of the people (the proletarian working class of a society and its leadership). And only when social-democratic surrogates are in place all around the world will social democracy be safe and peace possible.[10]

Thus, defensive war is any war conducted to protect a social-democratic homeland against foreign and domestic enemies—and must be maintained as long as even one enemy remains standing. Once a defensive war has accomplished its putative objectives, peace is possible. Democracy is the dictatorship of the leadership of the proletariat (democratic centralism). And morality is what "serves to [help] destroy the old exploiting society and . . . build up a new communist society."[11] Moreover, in a security environment in which words, ideas, perceptions, and so forth are important instruments of power and statecraft, leaders and the public must understand that the meaning of most of the Western political vocabulary has been dialectically distorted to the extent that, as in *Alice in Wonderland,* "words mean what I want them to mean"—that is, they are modified to suit the need and the cause at the time. The implications for meaningful dialogue and negotiations with an adversary seeking social "peace" are at best grim.

The Revolutionary Process

Revolution is not an event; it is a process. The operationalization of Lenin's scheme for conducting a defensive war against bourgeois enemies and compelling radical change begins with creating (and organizing, training, and employing) a body of experienced agitators (propaganda-agitator [or "agi-prop"] gangs, cells, groups, popular militias, or youth leagues).[12] In that connection, Lenin identified four general phases (stages or steps) in the revolutionary process: (1) a preparatory organizational period, preceding (2) the application of indirect voluntary-coercive methods, which are to be followed by (3) less-subtle repression and terror; and, if necessary to complete the process, (4) the obvious threat or direct use of military power. The agi-prop gangs have roles to play in all four phases of the revolutionary process and—with the possible direct use of military power—provide a systematic and holistic approach to revolutionary conflict.[13]

The tasks of these state-supported gangs include, first, "spreading, by propaganda, among the workers a proper understanding of the present social and economic system . . . and an understanding of the struggle between classes." Second, and inseparably connected with propaganda, is agitation. "Agitation means taking part in all manifestations of the working-class struggle.... There is NO issue affecting the life of workers . . . that can be left unused for the purposes of agitation."[14] Lenin further argued that if this instrument of statecraft succeeds in tearing apart the fabric on which a targeted society rests, then the resulting violence and destabilization can serve as "the midwife of a new social order."[15] Thus, liberal democracies all around the globe face total, painful, and long wars against the consensual basis of their political-economic-social systems.

The principal tools (methods/means) of the propaganda-agitator gangs include public diplomacy at home and abroad; intelligence, information and disinformation, and propaganda operations; cultural and political manipulation measures (such as bribery, corruption, coercion, repression, terror, subversion, and cyber war); and specifically prescribed covert and overt violence (instigating demonstrations, strikes, riots, and other civic violence, in addition to mutilation, murder, kidnapping, arson, and other persuasive intimidation actions).[16] The intent is for the propaganda-agitator gangs to act in such a manner as to evade foreign and internal notice and commentary and, as a result, allowing the supporting state to avoid responsibility for its actions.[17] But if all these indirect soft-power efforts fail, Lenin insisted that there is another—a final, decisive, and direct hard-power—instrument of statecraft: the armed forces and their "internationalist and liberating mission."[18]

Examples of these indirect methods include but are not limited to donating money, personnel, and other assets to cooperative officials, candidates, trade unions, political parties, and parliaments; withdrawing assets from uncooperative individuals, unions, parties, and so forth; establishing control of key businesses to influence or manipulate a given economy; making personal visits and phone calls to people one wants to influence or manipulate; publicly questioning officials, policies, and practices through statements, interviews, and articles in official and semiofficial or accommodating media outlets; publicly and officially denouncing and perhaps criminalizing or demonizing individuals, factions, parties, and other uncooperative entities; initiating sanctions against, or interfering with, the interstate and intrastate flow of trade, transportation, energy, and other commerce; rigging or buying elections at local and national levels; disseminating derisive information, disinformation, and propaganda against uncooperative or hostile individuals, parties, and institutions; mobilizing demonstrations, strikes, civil violence, riots, and other types of agitation in targeted localities, regions, and countries; and, again, as noted above, conducting carefully targeted mutilations, kidnappings, assassinations, and other persuasive-coercive and terrorist-repressive activities.[19]

There is nothing new or surprising in the above list of agi-prop methods or tactics, and they are not the exclusive property of contemporary Leninists but instead are well known and have been commonly used in national and international statecraft for at least 2,500 years. The universal historical reality of Lenin's revolutionary process (or "protection of Russia") goes way beyond the scope of this chapter. Likewise, the historical reality of Lenin's revolutionary process can be seen in a long list of Soviet—and now Russian—applications of these methods over the period from 1917 to the present. That too is beyond the scope of this chapter. Here, the discussion must necessarily be confined to background information regarding the formation of Nashi and a few selected situations that represent recent historical reality and have been credited to that organization.

RECENT HISTORICAL REALITY CREDITED TO NASHI

As early as 1897, in his pamphlet entitled "The Tasks of the Russian Social-Democrats," Lenin emphasized the key importance of the organization, training, and utilization of propaganda-agitator groups.[20] Then, in a speech to the Third All-Russian Congress of the Russian Young Communist League (Komsomol) in 1920, Lenin elaborated on the tasks that the youth leagues were expected to perform: "The young communist league will justify its name only when every step in its teaching, training, and education is linked up with participation in the common struggle of all working people against the exploiters."[21]

Thus, the Komsomol had been organized and was functioning by 1920. Oversight of such a broad mandate was provided by the Comintern (Communist International), operating in tandem with the secret and special services, the Communist Party, and the military. Later, after 1943, it became a primary instrument of ideological warfare and operated under the direction of the International Department (ID) of the party's Central Committee. The organizations and agents they manipulated were steadily expanded in range, scope, and number. Sovietologist Jan Adams has commented that the youth league's activities in the 1980s were so subtle, quiet, benign, and incremental that they largely eluded foreign commentary.[22] Komsomol's primary tasks up to that time were to proliferate client and communist states; cultivate the relationships of the Communist Party of the Soviet Union (CPSU) with nonruling communist parties, socialist-oriented parties, and national liberation movements; and expand the Soviet Communist Party's power and influence throughout the world.[23] The strategic objective of this kind of effort was straightforward. Increasingly, movements claiming to be states in embryo (national liberation movements) or defending "allied" state and nonstate actors act as challengers to the stability of a given state and can indeed play the role of Lenin's "midwife" to new social systems.

Most recently, after 2001, the Russian youth organization was resurrected, reconstituted, renamed Nashi, and placed under the direct control of the president of the country, outside any accountable structures. When Nashi came into public view in 2005, its primary task was to defend Russia (in actuality, President Vladimir Putin) from any internal or external attempt to generate a revolution and to thereby allow Putin to continue to control the Russian government. This task was articulated in somewhat different terms, however: Nashi's mandate was to "defend Russia against a Western plot to encircle the country."[24] This mandate stemmed from well-organized youth demonstrations in Ukraine that forced the country's pro-Russian president to hold free and fair elections that his West-leaning opponent won. Reportedly, Russia's President Putin was not amused and created his own youth organizations, Nashi and the Young Guard (the youth wing of the dominant political party in Russia, the United Russian Party [YeR]). Putin's intent was to create counterfoils for youth groups being organized by his Russian opposition and to ensure the succession of his regime in the 2007–2008 election cycle.[25] Accordingly, Nashi has been characterized as an eclectic group of thugs and college students who have taken Russian politics into the streets.[26] Nashi is also considered a de facto army of civilian zealots and a constant wild card in the "Defense of Russia."[27]

Nashi has not only played a part in the internal defense of Russia but also may have had a critical role in two external cases: Estonia in 2007 and Georgia in 2008. These are examples of the implementation of the Nashi Manifesto, issued in 2007: "We acknowledge that the struggle for military dominance is a dead end for mankind. Mankind wants to live in peace. Although we do not intend to impose our will by force, we will not allow Russia to be weak. . . . If our movement consists only of declarations, it will be forgotten before it dissolves. But, this is not our path. We are a tough squad, and we will demonstrate our solidarity by defending every single one of our allies and taking to the streets, not through abstractions."[28]

The Direction Nashi's Internal Path Follows

Succession in Russia, for which there is no established legal procedure, leads many scholars and observers to the conclusion that the system resembles a feudal society and state. The basis of Russian political affiliation, similar to that of tsarist Russia, remains the faction or "clan," political patronage, and clientelism. This so-called Moscovite Model is based on the tsar's patrimonial power over the national economy, including the institution of the boyer service state, in which all must serve the tsar to acquire the rents they seek. As a consequence, tsars and presidents deliberately encourage the division of the elites into rival and competing factions. As one example, in Russia there are multiple military and security organizations. Because of zero-sum game–type competition, no two or more organizations are likely to cooperate with each other, and no single organization is powerful enough to challenge the head of state. The tsar or president, however, directly controls each of the various youth groups and military and security organizations and can manipulate one or another or all for his own purposes. The elites' competitive struggle for access, power, and property is constant, never ending—and manipulated by whoever is taking the role of the tsar.[29]

Every post-Soviet succession has been accompanied by force, electoral fraud, and a steadily narrowing of democratic and public political participation. The most recent succession struggle was no different. It featured gross electoral manipulation, arrests of high-ranking public officials and military officers, and murders of well-known journalists. These actions were also accompanied by large-scale transfers of property to one or another clan. Among these factions are groups such as Nashi, which is known to have mobilized public opinion on behalf of the Putin regime and against its domestic and foreign critics. What we see here is a comprehensive strategy of internal consolidation of political power, coupled with the portrayal of Russia as a besieged fortress threatened by internal and external enemies. We also see political and organizational moves that can only lead to a police or neo-tsarist state.[30]

A closer examination of Nashi activities in the period leading up to the elections of 2007–2008 discloses a powerful political instrument in the form of harassment of "nonpatriotic" elements involved in the Russian political process and of foreign countries that had particularly displeased the Kremlin. The organization accomplished this with a range of different methods: (1) conducting paramilitary training in preparation for challenging internal and external opposition; (2) conducting ideological instruction and the promotion of Russian nationalism, including systematic anti-Western indoctrination; (3) organizing riots and demonstrations against foreign and domestic opponents for actions that affronted the Kremlin; and (4) playing a strong role in ensuring the succession of the Putin regime.[31] All these activities, while seemingly quite different, are highly interrelated and have the same political objective, that is, all contribute directly or indirectly to the "defense of Russia."

Paramilitary Training. Every summer since 2005, Nashi has established recruiting camps all across Russia. New members watch propaganda films, listen to lectures by top politicians and bureaucrats, and participate in basic military training courses, physical tests, and leadership exercises.

Additionally, Nashi members have also been observed on assault courses, practicing field-stripping Kalashnikov rifles and Makarov pistols, and target shooting. Nashi trains volunteer police troops who help regular police to patrol the streets—and if necessary, "beat hooligans." Reportedly, Nashi members can also join a group called SplaMeran that offers psychology courses for team leaders. The purpose of the psychology courses is to help leaders understand psychological and information (cyber and disinformation) wars and defend the principles of the government.

Much of this kind of activity is thought to be orchestrated by Vladislov Sarkov, President Putin's right-hand man for political and media issues, who meets regularly with Nashi leaders to organize propaganda campaigns and political demonstrations. The Kremlin thus appears to be grooming a militant youth movement as de facto enforcers of its nationalistic vision. In addition, well-qualified Nashi activists apparently are being groomed for government posts and election to the Duma.[32]

Ideological Instruction of the Russian People. Ideological introduction and anti-Western indoctrination are required as a result of the palpable official fear that independent civil society organizations might promote revolutions in Russia and other post-Soviet countries like that experienced in Ukraine in 2004. Putin's regime arguably needs to have an external enemy and domestic patriotism to maintain its popularity and find ways of legitimizing itself. The main problem and threat is the perceived Western desire to weaken, exploit, and rob Russia. Vladimir Putin, in these terms, is presented as a bulwark of Russian patriotism and the only leader capable of confronting the United States' intervention in Russia and protecting what is left of the imperial/Soviet heritage.[33] Three specific examples of Nashi ideological instruction and anti-Western indoctrination follow:

> In July 2006, Nashi demonstrated strong official displeasure with British ambassador Anthony Brenton's actions and made a public example of him. His offense was that he had spoken at a conference of an opposition party, Another Russia. He was

subsequently accused of helping to fund that organization and conducting a British spying operation out of his offices. The ambassador was repeatedly followed and harassed by Nashi activists in a prolonged, very open and bitter campaign of intimidation that lasted for over four months, until the British Foreign Office finally replaced him.[34]

In April 2007, Nashi deployed 15,000 volunteers throughout Moscow to hand out brochures and special cards for mobile phones to prepare people to receive instructions regarding what to do in the event of a pro-Western revolution.[35]

Prior to the 2007–2008 election cycle, Nashi issued a pamphlet identifying the organization's "Gallery of Traitors." The pamphlet featured twisted portraits of opposition political leaders, who were declared "enemies of the people." These individuals were denounced as scheming to subvert Russia and turn it over to foreign spies and conspirators.[36]

Demonstrations. At first glance, there appears to be little difference between the analytical concept of ideological instruction and indoctrination and that of demonstrations. The difference, however, is considerable. In this time when images, ideas, and perceptions are important means of influencing public opinion and leaders' decision making, demonstrations are an integral part of that equation. Demonstrations are powerful means of creating perceptions in the public mind. In these terms, demonstrations are an instrument by which to convey images and perceptions and to inform, instruct, and indoctrinate both an internal and an external target audience.

When the Kremlin and Nashi come together to decide what message they want to transmit to a given audience, they must also decide on the instrument they want to employ. Thus, ideology or doctrine is the message. A demonstration is a vivid instrumental means of sending that message. Nashi has conducted several effective demonstrations, beginning in 2005, when the organization garnered patriotic and positive national and international publicity after 50,000 activists demonstrated on May 5 (Victory Day [VE-Day] in Europe in World War II); the demonstrators were honoring the Russian victory, demanding respect for veterans of World War II, and opposed to evading the national draft and against "skinheads."[37]

The following year, when the governor of the Parm region (appointed by Putin) allowed a member of the opposition to attend a youth conference, hundreds of Nashi demonstrators picketed his offices. They demanded that he apologize for his irresponsible and unpatriotic act. The governor promptly did that, reportedly because he read the demonstration as the Kremlin's dissatisfaction with his administration.[38]

Several other demonstrations occurred in 2007. When Estonians removed a statue of a Russian soldier from the main square in Tallinn (placed there during the Russian occupation), a Nashi-led mob shut down the highway between Russia and Estonia. They had one message: "You are driving toward Fascist Estonia." In Moscow, Nashi demonstrators stormed a press conference being conducted by the Estonian ambassador. They retreated only after bodyguards sprayed the Nashi activists with pepper gas. At the same time, and not likely by accident, President Putin commemorated the Soviet victory over Nazi Germany with a massive military parade and dark warnings of "new threats to world security, as during the time of the Third Reich."[39]

One other notable demonstration occurred on March 3, 2008: Nashi demonstrated in front of the U.S. Embassy in Moscow. They were protesting "Many U.S. Plots" against Russia and Western meddling in Russian affairs.[40]

Succession of the Putin Regime. As discussed above, everything in the various Nashi activities leading up to the 2007–2008 elections, which culminated in Demitri Medvedev's appointment as Vladimir Putin's successor and Putin's return to high office as prime minister, indicate that those elections would not and could not be free and fair. Indeed, the entire succession process, complete with arrests and murders (as noted earlier),[41] illustrates the continuing propensity toward unrestricted political warfare and suggests that Russia's democratic foundations are becoming less and less solid. Nashi's role in all that was not insignificant.

As one example, former World Chess Champion Gary Kasparov tried to mount a campaign in opposition to the Putin succession. Early in the election campaign, in 2005, Kasparov saw his tour of Russian's southern regions disrupted by power shortages, technical problems with audio equipment, and egg-throwing protesters. As a consequence of these Nashi activities, he was unable to speak in most of the cities on his tour.[42] When Mr. Kasparov arrived at the Moscow airport to go to Samara, he was told that his tickets did not appear to be correct, and he was not allowed to board the aircraft. The police took his passport and returned it only after the last flight to Samara had left. Meanwhile, Nashi activists dressed in costumes depicting psychiatrists distributed leaflets ridiculing Kasparov's political "diagnosis."[43]

Additional examples of Nashi activity that had a decided impact on the outcome of the elections include the bussing of migrants from Central Asia to locations all around Moscow, enabling them to cast multiple votes during the 2008 elections.[44] (Interestingly, Nashi was reported to have had 100,000 poll watchers at those elections.)[45] And on December 6, 2008, 30,000 red-clad Nashi activists and 900 policemen blocked traffic in Moscow in an end-of-election demonstration. Children ranging in age from eight to fifteen held up signs with slogans such as "Thank you, Putin, for stability in our future."[46]

Nashi's Possible Role in the External Patriotic Defense of Russia: Estonia (2007) and Georgia (2008)

For years, military experts and computer scientists have speculated about the possibility of a nation's infrastructure being attacked using computers rather than bombs or other military means. In recent years, such concerns have been heightened by writers such as Qiao Liang and Wang Xiangsui and Jorge Verstrynge. The two Chinese colonels, Qiao and Wang, set the theoretical basis of such activity,[47] while Verstrynge broadened the methods of contemporary asymmetric war to include a variety of methods of operation such as "divine surprises."[48] These unconventional ideas came to reality with large-scale surprise cyber-attacks on Estonia in 2007 and cyber-attacks followed by conventional military attacks against Georgia in 2008.[49] Even though classified as nonlethal, the cyber-attacks on Estonia impaired the infrastructure, stability, and national well-being of that country as effectively as a conventional large-scale bombing campaign would have done.[50] The attacks on Georgia and the annexation of South Ossetia and other territories did exactly what they were intended to do; that is, they succeeded in forcing Georgia and its Western friends to acquiesce to Russian will.[51]

The theoretical perspective offered by Qiao and Wang as early as 1999 is odious:

> If the attacking side secretly musters large amounts of capital without the enemy nation being aware of this, and launches a sneak attack against his financial markets, then after causing a financial crisis, buries a computer virus and hacker detachment in the opponent's computer system in advance, while at the same time carrying out a network attack against the enemy so that the civilian electricity network, traffic dispatching network, financial transaction network, telephone communications network, and mass media network are completely paralyzed, they will cause the enemy nation to fall into social panic, street riots, and a political crisis. There is finally the forceful bearing down by the army, and military means are utilized in gradual stages until the enemy is forced to sign a dishonorable peace treaty.[52]

This type of unconventional warfare should not be considered to be a test of expertise in creating instability, conducting illegal violence, or achieving commercial, ideological, or moral satisfaction. Ultimately, it is an exercise in survival.

Verstrynge takes Qiao and Wang's speculation further down the theoretical path, exploring the idea of a "divine surprise," a key element in contemporary asymmetric war that lies outside the rules of war. A divine surprise is simple, unconventional, and can make the flesh creep (9/11 comes immediately to mind). Primarily, however, it depends on surprise, both in the method and in the desired ends. The basic philosophy is straightforward: the ends (objectives) are aimed at the profound destabilization of a targeted state. Ultimately, such destabilization would exchange the existing order of social classes, private property, social relations, internal and external politics, and the capitalist regime for a socialist society, economy, and polity. Consequently, this objective means total war. "It is a struggle without clemency that exacts the highest political tension."[53] Nevertheless, regardless of the methods used in modern asymmetric war, war is war, and given the changes in external appearance, it is still the means to compel the enemy to accept one's interests.[54] This threat is not abstract, it is real.

The divine surprise in Estonia consisted of cyber-attacks, which may be linked back to organized nonstate actors within Russia, such as Nashi. Nashi's manifesto discusses "taking to the streets," which does not necessarily mean "imposing [one's] will through [military] force" but may instead rely on nonkinetic methods of conflict.[55] Fortunately, as in Estonia, indirect nonkinetic methods of conflict do not always achieve their radical objectives. (Though, as in the case of Georgia, if various indirect, nonmilitary measures fail, the threat of overt violence is always present in the rhetoric and actions of socialist democracy.)[56]

Estonia, April 26–mid-May 2007. Journalists Owen Matthews and Anna Nemstova began their report on the divine surprise in Estonia as follows:

> The attacks came in waves, with military precision, hours after Estonia removed a World War II statue of a Soviet soldier from downtown Tallinn last month. Virtual war broke out. News agencies, banks, and Government offices were targeted in a blitzkrieg of spam—an onslaught of billions of e-mails—many apparently generated in Russia, that brought down servers and jammed bandwidths to bursting. As Estonia's famous digital based free markets and democracy buckled under the strain, top NATO Internet security experts rushed to construct defenses against the world's first massive cyber-strike by a superpower on a tiny and almost defenseless neighbor.[57]

A closer examination of the Russian actions in Estonia reveals that the intent was to destabilize the targeted state through the primary means of computers, which were used to shut down oil and natural gas shipments, rail service, electronic and telecommunications infrastructure, banks and businesses, media outlets, and Estonian political parties and government.[58]

In addition to the direct electronic and cyber-attacks, computers were employed to organize riots, violent demonstrations, and looting. But even though the country was completely paralyzed (except for the Russian-organized civil violence), Estonia did not fall into social panic and a political crisis. Estonians remained calm, resolute, and determined not to fall back under Russian control. Early on, government authorities and most citizens recognized the presence of civilian-clad Russian "special forces" (including Nashi) and understood what these groups were in Estonia to accomplish. Estonians are familiar with Russian military and special forces; even after the fall of the Berlin Wall and the political implosion of the Soviet Union, Russian troops remained in Estonia until 1994. Moreover, the hackers who were responsible for most of the electronic upheaval were traced back to Russia—even into the president's offices from which Nashi operates. Nevertheless, the Russian government denied any involvement or knowledge of those Estonian charges.[59]

Estonian authorities also contend that the Russian-organized violence could have been used as a pretext for a military intervention. Though Western audiences might consider such a threat assessment to be far-fetched, the Estonians did not. And Estonian authorities emphasize that the operations listed above represent something quite close to state-sponsored terrorism. Yet Russia did not use direct military ways and means to force Estonia to accept its political dominance. Authorities contend that Russia was surprised and disappointed by the lack of support for the civil violence in which Estonia's ethnic Russians were expected to participate. Russia was also surprised by the quickness and resolution demonstrated by the European Union (EU), the Council of Europe, and NATO in support of Estonia against Russia.[60]

Thus, even though the Russian experiment with cyber war in Estonia probably did not go exactly as planned, it is instructive. The combination of information warfare (computer war or cyber-attack), gang penetration, incitement of diasporas or other similar ethnic congregations, incapacitation of politicians and political institutions, and the shutdown of national infrastructure represents a nonlethal but destructive form of warfare. Each, some, or all of the activities associated with information warfare can serve as elements of a paradigm for what is increasingly called asymmetric war. Such tactics are surrogates for military dominance that "is a dead-end for mankind."[61] Contemporary Russian leaders see information warfare in a similar light. Shortly after the Estonian experience, the deputy premier and former defense minister Sergie Ivanov stated, "The development of information technology has resulted in information itself turning into a certain kind of weapon. It is a weapon that allows us to carry out would-be military actions in practically any theater of war, and most importantly, without using military power."[62] Ivanov might have added that these activities, conducted by covert protagonists, are "plausibly deniable."

Georgia, July–August 2008. Journalist John Markoff, writing in the *New York Times,* documented cyber-attacks against Georgia's Internet infrastructure beginning as early as July 20, 2008. Those attacks overloaded and effectively shut down Georgian servers three weeks before the conventional "shooting war" began. However, these cyber-attacks had far less impact on Georgia than Estonia. In Estonia, vital activities such as power, transportation, banking, and government services are tied to the Internet. This is not the case in Georgia, which ranks 74th out of 234 countries in terms of Internet addresses, behind Bangladesh, Nigeria, Bolivia, and El Salvador. Nevertheless, Shadowserver, a volunteer group that tracks malicious network activity, reported that the cyber-attacks against Georgia spread to computers throughout the government before Russian air strikes began and shortly after Russian troops entered the separatist Georgian province of South Ossetia.[63]

Again, exactly who was behind these cyber-attacks is not known, and again, the Russian government said it was not involved. Yet numerous U.S. computer security researchers reported that they saw clear evidence of the involvement of a St. Petersburg group called the Russian Business Network (RBN).[64] That organization might or might not have been connected to Nashi, but one must always remember the diplomatic dictum of "plausible deniability." The coordination and timing of the cyber-activity appears to be anything but coincidental.

Regardless of the identity or intent of the perpetrators, the cyberattacks failed to achieve the desired objective, that is, the attacks failed to make the Georgian government and armed forces inoperable. Thus, an ill-considered Georgian attack during the night of August 7 into South Ossetia brought substantial Russian military forces streaming into Georgia. It was a complex, relatively conventional combined arms operation, and by August 12, the war was over.[65]

The original intent of the cyber operations appeared to be to paralyze Georgia and "cause the enemy nation to fall into social panic, and a political crisis."[66] That did not happen precisely the way Defense Minister Ivanov might have predicted. As a result, there was "the [inevitable] forceful bearing down by the Army" until Georgia and its Western friends were forced to bend to Russian will.[67] Additionally, Vladimir Putin appears to have drawn a line limiting NATO expansion and sending a strong message to Georgia, Ukraine, and the West that Russia is initiating a new, more assertive, and more imperial foreign and defense policy.[68]

The implications for the West and Russia's neighbors are sobering. The messy breakup of the Soviet Union left millions of ethnic Russians stranded in other post-Soviet states. Ukraine, for example, is 17 percent ethnic Russian. There are signs that the Kremlin is systematically reaching out to these various Russian-speaking communities (possible Trojan horses) through a range of programs to enhance Russia's nonkinetic power in areas once controlled by the Soviet Union and, earlier, by Imperial Russia. One program aims to "shape a new generation of highly nationalistic Russian politicians" out of youth activists from groups such as Nashi. At the same time, these programs are likely to work toward the cultivation and support of populists and neopopulists, national liberation movements, and New Socialist political parties and, in turn, to increase Russia's power and influence throughout the world.[69]

Where Else Nashi Leads

The succession that elected Medvedev and Putin to continue Putin's policies and programs was characterized by an intense, no-holds-barred competition for the rewards of power. Nashi played its part in mobilizing public opinion on behalf of the regime and in the patriotic defense of Russia. The resort to subtle and not-so-subtle violence, as well as the obvious adoption of the medieval "political continuism model" (a process in which the outgoing leader's protégé [or even wife] is chosen to be his successor), reinforces Russia's paternalistic and patrimonial traditions. Violence and continuism also enhance the role of those special groups that help make a succession possible. At the same time, those special services, including Nashi, help intensify rivalry with the West. In this model, the government that wants to succeed itself needs external and internal enemies, no matter how innocuous, to generate voter popularity and a certain moral legitimacy. Thus, if Nashi continues to play its role in helping provide the political needs of the governing elite in terms of the defense of Russia, there will be plenty of plausibly deniable work—and rewards.

Fifty years before Nashi was organized, Albert Camus described the ultimate path that would be taken by Nashi activists and people like them, those whom he says "have turned their backs on the present. They forget the present for the future, the fate of humanity for the delusion of power, the misery of the slums for the mirage of the eternal city, ordinary justice for an empty promised land. They despair of personal freedom and dream of a strange freedom of the species; reject solitary death and give the name of immortality to a vast collective agony."[70]

KEY POINTS AND LESSONS

- The most recent of the Russian youth organizations, Nashi emerged with the intent of building metaphorical Trojan horses, generating divine surprises, and influencing public opinion at home and abroad.
- The primary task of Nashi was and is to defend Russia from an internal or external attempt to generate a revolution, while allowing Putin to continue to control the Russian government.
- Nashi demonstrates how a state-supported set of "gangs" might fit into a holistic state or nonstate actor's effort to compel radical

political-economic-social change, as well as how traditional political-military objectives might be achieved indirectly rather than directly.
- Strategic leaders must understand that an enemy cannot only be the military formations a state or nonstate actor may be able to put into the field; rather, the enemy is now everyone and anyone who supports the "bourgeois class," or its equivalent, in any way. As a consequence, the conventional state-centric center of gravity can no longer be state or easily identified state-supported military or paramilitary formations. Instead, it must include external and internal decision-making leadership and public opinion.
- The principal tools (methods/means/instruments) of the Nashi-type propaganda-agitator organizations include public diplomacy at home and abroad; intelligence, information and disinformation, and propaganda operations; cultural and political manipulation measures; and overt and covert violence.
- In the run-up to the elections of 2007–2008, Nashi served as a powerful political instrument through internal tasks such as paramilitary training; ideological instruction and anti-Western indoctrination; and riots and demonstrations against foreign and domestic opponents.
- Nashi's external tasks appear to include plausibly deniable involvement in information war and organization of riots, demonstrations, and looting. These are weapons that allow an aggressor to carry out would-be military actions anywhere in the world, without using traditional military power.
- Nashi, then, appears to be a powerful instrument of nonkinetic power abroad, as well as an instrument for generating public opinion and influencing decision and policy makers at home.

War is changing. It is no longer limited to using military violence to bring about fundamental and radical political change. Rather, a combination of means, including state-supported gangs such as Nashi, must be used to co-opt, urge, and compel an adversary to do one's will. Neither cyber-attacks nor superior firepower is a panacea, and technology may not give one a knowledge or information advantage. The astute modern warrior will tailor his campaign to the adversary's political-economic-psychological-military vulnerabilities and to the adversary's political-psychological perceptions. This represents a sea change in warfare and requires nothing less than a paradigm change in how conflict is conceived and managed.

B.H. Liddell-Hart reminds us that the effectiveness of armies depends on the development of new methods that "aim at permeating and dominating areas rather than capturing lines; at the practicable objectives of paralyzing the enemy's action rather than the theoretical aim of crushing his forces. Fluidity of force may succeed where a concentration of force merely entails a perilous rigidity. . . . Thus, the concept of 'cold war' is now out of date, and should be superseded by that of 'camouflaged war.'"[71]

CHAPTER 5

GUATEMALA AT RISK

Drugs, Thugs, and Radical Political Change

The market-oriented economic reforms of the 1980s and 1990s, combined with a few years of commodity-driven prosperity, are starting to create societies in Latin America that are more wealthy and less unequal. Guatemala (although not anywhere near the top of the list of such societies) is a case in point; nevertheless, according to *Latinobarómetro*, only 8 percent of Guatemalans think that democracy works better in their country than in the rest of Latin America. This is the lowest figure in the region. At the same time, perceptions of citizen insecurity are worse in Guatemala than anywhere else in the hemisphere. Thus, despite a richer, more stable economy in Latin America—and Guatemala—there are still a few things "to put right" before Guatemala can build successfully on recent economic growth.[1]

There are at least three big concerns regarding threats to contemporary security and stability (and sovereignty) in Latin America in general, and Guatemala in particular: widespread crime and violence, much of it perpetrated by criminal gangs and other transnational criminal organizations (TCOs);[2] criminal violence that goes hand-in-hand with political inaction, instability, poor policy choices, weak political and security institutions, and the undermining of the rule of law;[3] and corruption of "the authorities themselves, who rarely investigate crimes, but demand bribes."[4] The political-economic-social disequilibrium caused by these destabilizers generates the correlation of forces that make for a revolutionary situation. These same destabilizers are the underpinnings of John Holloway's notion of attaining power and precipitating radical political change.[5]

In the Guatemalan situation, the question is: Who will achieve that radical change—a narco-criminal or a New Left revolutionary element? Regardless of the identity of the nonstate actor, we look at the instability being generated by criminal violence, weak institutions, and corruption, coupled with the possibilities for radical political change in Guatemala. And we wonder if anyone else is watching, or cares. We sincerely hope so.

THE HISTORICAL-POLITICAL CONTEXT

In its existence as an independent state since 1838, Guatemala has experienced generally weak predatory dictatorships that have been tools of, first, the big landowners, then, progressively, an agricultural–industrial–foreign investor elite and an agricultural–industrial–foreign investor–military elite. Now the elite consists of remnants of the old agro-industrial–foreign investor elite along with foreign interests that range from the United Nations (UN) to private nongovernmental organizations (NGOs) to former insurgents to narco-criminal organizations.[6]

Within the entire period from 1838 to the present, Guatemala has had to deal with only two aberrations. First, there was the Communist revolution of 1950. The Soviet-backed Jacobo Arbenz government was deposed in June 1954, and Arbenz's successor, Castillo Armas, was assassinated in March 1956. Armas had been accused of supplying arms to the "communist" insurgent opposition.[7] Second, Guatemala's more recent history has been overshadowed by the country's civil war, in which two major but related conflicts developed into a thirty-six-year civil war perpetrated by four different but related insurgent groups.[8] The first conflict began in the 1960s in the eastern part of the country. The second, and most brutal and bloody, conflict spread into the indigenous western highlands in the late 1970s and 1980s. Like the insurgents of the revolution of 1950, these guerrillas (insurgents) were also marked as communists and emissaries of Fidel Castro.[9] Whatever the labels, however, the umbrella organization that was intended to coordinate their separate activities (Guatemalan National Revolutionary Unity [URNG]) called for popular revolution.[10] The civil war finally ended with the United Nations–sponsored Peace Accords of 1996, which granted power to an entirely new set of controlling elites in the Guatemalan political culture.[11]

Background

According to historian Hal Brands, Guatemala has often been governed by an authoritarian state but never by a strong or effective government.[12] The creation of a new set of elites is key to understanding the weak state that emerged out of the 1996 Peace Accords.[13] These political actors are many and varied and can been seen in the following groupings.

First, the accords introduced international commissions from the United Nations into the Guatemalan political process. The intent was for the UN commissions to facilitate the implementation of the accords with the help of private nongovernmental organizations. These international actors have become important but not absolutely controlling influences in Guatemalan politics. Nevertheless, these international actors have been instrumental in

- Dismantling the army as an effective institution. Its size was reduced from 60,000 in the mid-1980s to 15,000 in 2008. It lost most of its share of the national budget and has been prohibited by law from engaging in domestic (internal) security activities.[14]
- Supporting a "bucolic vision" of the indigenous peoples of Guatemala and championing their civil rights, cultural recognition, protection, and relative political independence.[15]
- Supporting the former insurgent combatants (*ex-combatantes*) of the civil war period in their radical political agenda.[16] This, along with the great socioeconomic distance between the masses and the middle and upper classes, has encouraged a culture of impunity.[17]

Second, as Santiago Fernandez, Edward Fischer, and Hal Brands all argue, the UN commissions (primarily the International Commission against Impunity in Guatemala [CICIG]), the NGOs, and the former insurgents exercise more influence than the agro-industrial-military elites of the past.[18] They further argue that, as a result of the vacuum left by the removal of the army from the controlling elite, another elite has emerged since the signing of the 1996 Peace Accords: a criminal element associated with narco-trafficking. Hal Brands specifically states that the result of the civil war and the subsequent accords "weakened the only institution capable of maintaining some semblance of public order . . . leaving the field to opportunistic elements that would prey on [the traditional state] weaknesses."[19]

Thus, a weak Guatemalan state prone to disorder and violence has become a focal point of the international illegal drug trade, creating lucrative opportunities for a wide range of criminal elements that operate in at least seventeen of the twenty-two departments of the country.[20] Both Brands and Fernandez organize these elements into three groups.

Transnational Criminal Organizations (TCOs). These include Colombian and Mexican cartels and private armies such as the Mexican Zetas operating in Guatemala.[21]

Pandillas (Gangs) and Maras. Maras (short for a specific gang, the Mara Salvatrucha) and *pandillas* are at the center of the crime epidemic afflicting Guatemala: the pandillas are involved in petty extortion, robbery, small-scale drug trafficking, kidnapping, and murder; the Maras are involved in all these activities and more, including drug trafficking, human trafficking, arms trafficking, stolen car trafficking, and racketeering. At the same time, the Maras act as mercenaries for the TCOs and "hidden powers" (noted below) by providing security for drug and arms shipments and distributing illegal drugs in the Central and North American markets.[22]

The "Hidden Powers." These consist of corrupted prominent businessmen, current and former military and police officers, politicians, and bureaucrats. In other words, the hidden powers are integral elements of Guatemala's socioeconomic elites.[23]

As a consequence of the addition of these criminal elements into the Guatemalan political process over the past fifteen years, there has been an increasing level of public insecurity. Guatemalans are turning away from the ineffective state and turning toward private security organizations. Ironically, some (or much) of the country's private security is being provided by the hidden powers. Additionally, these elites have insinuated themselves into government and society and have further diluted the general weakness of the state, to the point where the president appears not to see, hear, or do anything.[24]

Ramifications

Guatemalan security, democracy, and effective sovereignty are being impinged upon every day of the year. Security is virtually nonexistent because most of the national territory is under the control of TCOs, private armies, and gangs. Elections are meaningless because the electoral process is controlled by the traditional elites in alliance with gangs and transnational criminal organizations (that is, the hidden powers),[25] in addition to the radical left, as represented by the ex-combatantes. Sovereignty is a joke because the Guatemalan state cannot control the national territory and the people in it. Also, crediting Guatemala as a sovereign state is difficult when the TCOs and gangs hold a veto over responsible governance and public well-being.[26] The state is unable or unwilling to provide the rule of law, because of weak institutions, weak political parties, the penetration of the state by criminal networks, and the accompanying impunity and political inaction. Personal and collective insecurity, all-pervasive impunity for lawlessness, and political inaction are considered the country's main problems. No one can or will deal with them. These problems have created a crisis of a dimension and magnitude unequalled in Guatemala's 175-year history. That crisis takes us back to where we began. From 1838 to the present, Guatemala's weakly led governments have generally remained in the service of the controlling agricultural-industrial elites, coupled with foreign interests (to include transnational criminal networks and former insurgents).[27]

WHAT CRIMINAL VIOLENCE, POLITICAL INACTION, AND CORRUPTION HAVE WROUGHT IN GUATEMALA: A FEW EXAMPLES

The following short sketches are intended to help civilian and military leaders, opinion makers, academics, and the concerned public understand the basic reality of contemporary Guatemalan politics. That country, however, is not unique. The problems illustrated are epidemic all around the world. They serve as examples to explain how and why politics becomes caprice or whim that is unpredictable, erratic, and impulsive, and how and why some political leaders appear to have neither good nor bad character but, rather, no character. These examples also serve to help explain how and why criminal violence, corruption, and political inaction lead a country into the process of state failure. These are issues that strategic leaders must think about and deal with effectively. Otherwise, they can only watch the global security arena become further engulfed in a political-economic-social-military-chaos generated by criminals, insurgents, and corrupt businessmen, bureaucrats, and politicians—to say nothing of demagogues and hatemongers.[28]

Elections in a Traditional Mayan Community, 2007

Election time in Guatemala means rallies and banners—and body bags. In a traditional Mayan town such as Santiago Atitlan, on the southern shore of Lake Atitlan, there are two ways to win an election. Campaigning is one way, though the odds of success from this approach are not particularly good. The other is to call on Maximon. A quasi-deity carved from wood, Maximon is located in a darkened room near the town center. He is ringed by candles and doused with rum. The figure is said to have magical powers—including the ability to sway an election.

Before the national elections in September 2007 that brought Alvaro Colom to the presidency, several representatives of various candidates for local and national offices called on Maximon to provide him with shot or two of rum and ask for his help in the election. In addition to that type of campaigning, it was a violent campaign. At least fifty people tied to various campaigns for various offices were reported killed in Santiago Atitlan and its surrounding area. When asked who Maximon's keepers were, no one knew. No one ever does.

As one example of Maximon's powers, Rigoberta Manchu, the first indigenous candidate for president of Guatemala and the winner of a 1992 Nobel Peace Prize for her advocacy on behalf of that country's marginalized Mayan community, won only 347 out of the 14,000 votes cast for president. She won fewer votes than former general Perez Molina, implicated in the brutal suppression of the Mayan community during the 1960–96 civil war. General Perez Molina was not at all popular in Santiago Atitlan, but he received over 2,000 more votes than Ms. Manchu. Alvaro Colom, miraculously, received over 11,000 votes in that election and went on to win the Guatemalan presidency.[29]

Activities in the "Protected" Maya Biosphere Reserve: Politics in the North of Guatemala

Activities in the Maya Biosphere Reserve in northern Guatemala are good examples of government failure to extend or maintain a legitimate sovereign presence throughout the national territory. This failure leaves a vacuum in which gangs (Guatemalan and Central American Maras) and private armies (the Mexican Zetas), drug cartels, and other transnational criminal organizations, such as the ex-combatantes, may all compete for power in that supposedly "protected" space. In that connection, such areas are not "lawless" or "ungoverned." Instead, these territories are governed by the drug barons, gangs, and warlords operating there, who have been known to pursue ambitious economic and political agendas. Those agendas clearly and unequivocally take the gangs, private armies, and insurgents into mercenary activities and intrastate war. Their objectives (which define war) are to (1) neutralize, control, depose, or replace an incumbent government; (2) control parts of a targeted country and create autonomous enclaves that are sometimes called criminal free-states, para-states, or virtual states; and (3) in doing so, radically change the traditional authoritative allocation of values (governance) to the values of the reigning criminal and/or insurgent leaders.

The Mayan Biosphere Reserve, Central America's largest "protected" area, spans about 20 percent of Guatemala. It is located in the "lawless" regions that border Mexico in the north and Belize in the east. For some time, it has been one of the major cocaine transporting routes and distribution centers from South America to the lucrative markets of the United States. It is also the place where Guatemala's President Alvaro Colom has said that he would like to create a major eco-tourism destination. But according to an American anthropologist working in the area, Richard D. Hansen, the Mayan Biosphere Reserve is a place where "all the bad guys are lined up to destroy the reserve. You can't imagine the devastation that is happening."[30]

The devastation of the region is being accomplished by an assortment of "bad guys" including cattle-ranching drug barons, looters, poachers, squatters, gangs, and other mercenary organizations. Looters moving freely in the reserve have made off with priceless items from Mayan graves, temples, and other ancient structures. Poachers are targeting rare species of jaguars, crocodiles, river turtles, spider monkeys, and scarlet macaws and are—along with the looters—destroying the habitat. Squatters are peasants who have come into the area in search of free farmland. But to be allowed to keep their newly cleared land, they have been "asked" to act as armed guards (pickets) to prevent or warn the cattle-ranching drug barons of intruding government officials and soldiers.

This takes us to the politically as well as ecologically destructive drug barons and their mercenary enforcer gangs. The cattle ranchers have been cutting and burning approximately 37,000 acres of forest a year since 1996 to create pastures for approximately 2.5 million cattle. These ranchers have also been using their newly cleared lands for airstrips, money laundering, continued drug and other illicit trafficking, and beginning the process of their legitimization in Guatemalan society. The gangs (Maras and Zetas) establish, protect, and control areas around the overland drug routes and airstrips in a region that has become a major distribution center for drug, arms, and human trafficking from South to North America. This narco-criminal free-state in the Maya Biosphere Reserve is a far cry from President Colom's vision of a lush Maya-themed vacationland.[31]

Problems in High Places

The 2007 electoral campaign that brought President Colom to power was a violent one. The main themes were development and peace. Yet the flurry of bullets and the occasional machete attack made the campaign the bloodiest in the history of Guatemala.

Two things made the bloodletting different this time. First, in the past, the traditional political violence occurred between one candidate and another, or their followers. This time, the violence was attributed to narcotics traffickers, paramilitary groups, gangs, and the hidden powers. None of these entities are trigger shy. As one example, sixty-one violent attacks against high-level candidates in the 2007 campaign were reported, including the murders of seven congressmen and several other candidates for national offices. With plenty of money to spend, drug traffickers and their allies finance as many electoral campaigns as they think necessary to guarantee a friendly reception when asking government officials for political favors. If ever there is resistance, it is normally met with gunfire. An analyst at the Guatemalan research group Security and Democracy, Iduvina Hernandez, argues that "controlling the political system is their goal."[32]

Second, of the more than three thousand murders on the books in 2009, one jolted Guatemala like no other: the death of a prominent lawyer, Rodrigo Rosenberg, reportedly sent hundreds of thousands of people into the streets to protest President Colom's alleged involvement in that murder. The act was not much different from virtually any other political murder in Guatemala. Mr. Rosenberg was shot by one assailant as he rode his bicycle near his home. The coup de grâce was administered by a second gunman. No one knows who these individuals were. They are, however, thought to be like hundreds of others who turn out to be hit men hired by some shadowy figure who is never identified.

What makes this situation different is that Rosenberg made a video a few days before he died. He said, "My name is Rodrigo Rosenberg Marzano, and unfortunately, if you are watching this message, it is because I was assassinated by President Alvaro Colom." In the video and the written statement that accompanied it, Rosenberg went on to say that the president and those around him were involved in a corruption scandal and the murder of one of Rosenberg's clients and his daughter. Rosenberg offered no proof of his allegations, but the fact that he foretold his own murder gave him considerable credibility.

Colom has denied having anything to do with the alleged killings. Additionally, the president turned the case over to the UN's International Commission Against Impunity in Guatemala. The Spanish jurist who headed the commission reportedly remarked, "This is like a John Grisham novel, but it is real. . . . This case is important because it affects the governability of the country."[33] Nevertheless, given Guatemala's history of high-profile murders that are never solved, some Guatemalans fear that this case will follow the familiar pattern of the past.[34]

Problems in Guatemala's Justice System

Under an unusual agreement with the United Nations, Guatemala has invited foreign prosecutors and investigators to help prepare delicate cases that might otherwise have been shelved by intimidated or corrupt officials. The responsible UN organization is known as the CICIG (the International Commission Against Impunity in Guatemala). The agreement was recently tested in a political struggle involving the unexplained death of a former minister of the interior, Vinicio Gomez, who was investigating drug trafficking near the Mexican border. Additionally (though not necessarily connected with the death of Mr. Gomez), a former president is in jail facing charges of embezzling money from the military budget, two former National Police chiefs have been arrested on drug trafficking charges, and six former Defense Ministry officials have been charged with fraud and embezzlement. At the same time, other government officials, including the security detail of another former National Police chief, were being investigated on drug charges and illegal child adoption schemes.

Controversy regarding these and other problems in the Guatemalan justice system led to the resignation of the UN-appointed director of CICIG, Spanish jurist Carlos Castresana. The trouble started at the end of May 2010 when President Colom appointed a new attorney general (prosecutor general), who began to remove prosecutors and investigators who had been working with CICIG. Mr. Castresana protested but was ignored. He resigned on June 7, after asserting that the new attorney general, Conrado Reyes, had links to organized crime. To keep the process going, the UN immediately appointed a new director for CICIG. (The problem here was that, under the terms of the agreement between the UN and Guatemala, both the UN director and the attorney general were required to proceed together in their duties regarding Guatemalan justice.) Then Guatemala's highest court removed Mr. Reyes from office. It ruled that the complex legal procedures for selecting a prosecutor general had not been followed by President Colom. In the meantime, the president was extremely slow in appointing a replacement for Reyes, and the country (and the UN) had to await presidential action.

Not surprisingly, CICIG's activities have failed to provide the results most Guatemalans expected. The commission has made steps toward shaking up the culture of impunity and strengthening the rule of law.[35] Nevertheless, according to the wife of the dead former interior minister, "The people in organized crime are very well structured. They have permeated all the institutions" of Guatemala.[36] A congresswoman, Nineth Montenegro, is a bit more explicit: "We have a police force that is penetrated. . . . We have a Prosecutor General's office that is penetrated [and cannot function without a new attorney general to work with the UN] . . . and we have a President who appears not to see anything."[37]

The Threat

The TCO–gang–hidden powers challenge to Guatemalan national security, stability, and sovereignty and the attempt to neutralize, control, or depose incumbent governmental institutions takes us to the strategic-level threat. In this context, crime, violence, corruption, impunity from prosecution, governmental inaction, and instability are only symptoms of the threat. The ultimate threat is either state failure or the imposition of a radical social-economic-political restructuring of the state and its governance in accordance with criminal values. In any case, the TCO–gang–hidden powers phenomenon contributes to the evolutionary state failure process by which the state loses the capacity and/or the will to perform its fundamental governance, service, and security functions. Over time, the weaknesses inherent in its inability to perform the business of the state are likely to lead to the erosion of state sovereignty (authority) and legitimacy. In the end, the state no longer controls either its territory or the people in it. In that connection, the result will be determined by the best-organized, best-motivated, best-disciplined, and best-armed political actor on the scene at the time.[38]

OUT OF THE SHADOWS OF THE PAST, IMAGES OF THE FUTURE: POSSIBLE DIRECTIONS FOR GUATEMALA

Criminals, corrupt politicians and government officials, and revolutionaries are talking about good times ahead. Members of the first three categories can see opportunities for further self-enrichment. Revolutionaries, including populists, twenty-first-century socialists, and former insurgents, sense the development of a political situation (the appropriate correlation of forces) in Guatemala that will allow them to realize their dreams.

One way or another, protagonists are seeking to destroy present structures to build new utopian political-economic-social systems. But again, the human and physical elements that perpetrate the destruction of present political-economic-social structures—crime and violence, corruption of authorities, and weak institutions and political inaction—are symptoms of state failure. Thus, state failure plays a major role in enabling radical political change in Guatemala.

The Issue of State Failure

State failure is an evolutionary process, not an outcome, and is often brought on by poor, irresponsible, and insensitive governance. State failure can be exacerbated by nonstate (criminal or insurgent) and other groups that, for whatever reason, want to take down or exercise illicit control over a given government. The narco–gang–insurgent–hidden powers nexus in Guatemala represents an unconventional, asymmetric threat to the authority of the central government. Through murder, kidnapping, corruption, intimidation, destruction of infrastructure, and other means of coercion and persuasion, these violent, internal nonstate actors compromise the exercise of state authority. The government and its institutions become progressively less and less capable of performing the tasks of governance, including exercising their fundamental personal security function of protecting citizens. As a result, the narco–gang–insurgent–hidden powers nexus becomes increasingly wealthy and powerful, and the country deteriorates further and further toward failed state status.[39]

Peru's Sendero Luminoso calls violent and destructive activities that facilitate the processes of state failure "armed propaganda." Drug cartels operating in that country and throughout the Andean Ridge of South America and elsewhere call these activities business incentives. Thus, in addition to helping to provide wider latitude to further their immediate objectives, Sendero's and other violent nonstate actors' armed propaganda and business incentives are aimed at lessening a regime's credibility and capability in terms of its ability and willingness to govern and develop its national territory and society. This type of debilitating and destabilizing activity generates the most dangerous long-term security challenge facing the global community today.[40]

More specifically, failing or failed states in Latin America, Africa, the Middle East, and Asia serve as breeding grounds for instability, insurgency, and terrorism. A breakdown in institutional governance can breed or exacerbate humanitarian disasters and major refugee flows. Such states can host networks of all kinds, including criminal business enterprises and/or some form of ideological, religious, or populist crusade. They also spawn a variety of pernicious and lethal activities and outcomes, including torture and murder; poverty, starvation, and disease; the recruitment and use of child soldiers; trafficking in men and women and in selling human organs for transplants; trafficking and proliferation of conventional weapons systems and weapons of mass destruction; genocide, ethnic cleansing, and warlordism; and criminal anarchy and insurgency. At the same time, these networks and activities normally are unconfined and spill over into regional syndromes of destabilization and conflict.[41]

Additionally, failing and failed states do not go away. Ample evidence demonstrates that failing and failed states become dysfunctional states, rogue states, criminal states, narco-states, or new people's democracies. Moreover, failing and failed states tend not to be interested in human rights or cooperate on shared problems such as illegal drugs, illicit arms flows, debilitating refugee flows, and potentially dangerous environmental problems. In short, the longer they persist, the more they and their associated problems endanger regional and global security, peace, and prosperity.[42]

At the same time, in the global security environment, international organizations and willing national powers are increasingly called on to respond to conflict generated by all kinds of instabilities and destabilizers. Accordingly, the international community increasingly is expected to provide the leverage to ensure that legitimate governance is given to responsible, incorrupt, and competent leadership that can and will address the political, economic, and social root causes that underlie a given destabilizing situation. This new "sovereignty of responsibility" concept has serious implications in terms of failing and failed states. As demonstrated in the Guatemalan case, the conscious positive or negative choices (or nonchoices) that a government takes about how to conduct national stability efforts will define the future of the state.[43]

The Problem of Narco-Corruption

Professor (former Ambassador) David C. Jordan would argue that Guatemala is an "anocratic democracy," that is, Guatemala is a state that has the procedural features of democracy (elections) but retains the characteristics of an autocracy. That autocracy faces no scrutiny or accountability. At the same time, Guatemala is a market state that is moving toward "criminal free-state" status. That is, Guatemala is a state in which political power is migrating from the state to small, nonstate actors who are organized into sprawling networks that maintain private armies (Zetas), enforcer gangs (the Maras), treasury and revenue resources, welfare services, and the ability to make alliances with state and nonstate actors and to radically change the Guatemalan political-economic-social system. This correlation of forces has generated an extremely violent, dangerous, and destructive internal security situation in Guatemala that has been all but ignored both inside and outside the country.[44]

The Anocratic Democracy. The policy-oriented definition of democracy that has been generally accepted and used in U.S. foreign policy, and throughout the world over the past several years, is best described as "procedural democracy." This definition tends to focus on the election of civilian political leadership and, perhaps, a relatively high level of popular participation on the part of the electorate. Thus, as long as a country is able to hold elections, it is considered a democracy—regardless of the level of transparency, accountability, resistance to corruption, ability to extract and distribute resources for national development, and protection of human rights, liberties, and security.

In contemporary Guatemala, we observe important paradoxes in this concept of democracy. Elections are held on a regular basis, but leaders, candidates, and bureaucrats are intimidated, corrupted, and/or assassinated before and after elections. Indirect threats, such as kidnapping and the use of relatively minor violence on a person and/or his family also play important roles before and after elections. Consequently, most Guatemalan elections cannot be described as genuinely "democratic" or "free." Moreover, crediting Guatemala as a democratic state is difficult as long as elected and appointed leaders are subject to the corrupting control and informal vetoes imposed by drug lords or other criminal nonstate actors (private armies or gangs). The final result of intimidation and coercion tends to erode the will and ability of the state to carry out its legitimizing functions.[45]

The Market State and the Gang-TCO Phenomenon. John Sullivan has identified an important shift in state form: from nation-state to market state, and from market state to criminal free-state status. As the ability to conduct violence devolves from traditional hierarchical state organizations to Internet-worked transnational nonstate actors, we can see the evolution of new conflict-generating entities (that is, small private armies) capable of challenging the stability, security, and sovereignty of traditional nation-states. These private entities (such as terrorists, warlords, drug cartels, enforcer gangs, criminal gangs, and ethnonationalistic extremists) respond to illicit market forces (such as illegal drugs, arms, and human trafficking) rather than the rule of law and are more than "stateless" or nonstate groups. They are powerful organizations that can challenge the rule of law and the sovereignty of a nation-state. They also are known to promulgate their own policy and laws—and impose their own values on societies or parts of societies, creating criminal free-zones and "badlands and bad neighborhoods" all around the world.[46]

In Guatemala, as an unintended consequence of devolving political power from the state to private nonstate entities, we see not only the erosion of democracy but also the erosion of the state. David Jordan argues that corruption at all levels is key to this problem and is a prime mover toward "narco-socialism."[47] Narco-politics has not only penetrated the executive, legislative, and judicial branches of the Guatemalan government but also goes down into the municipalities. The reality of corruption at any level of government favoring the TCO-gang phenomenon mitigates against responsible governance and public well-being. In these terms, the state's presence and authority is at best questionable. The corruption reality, then, brings into question the issue of effective state sovereignty. This is a feudal environment and defined by impunity, extreme violence, patronage, bribes, kickbacks, cronyism, ethnic exclusion, and personal whim. Some observers of the gang phenomenon, such as Phil Williams, assert that the coerced change toward criminal values in a society such as that in Guatemala leads to a "New Dark Age."[48]

A Possible New "Citizen's" Revolution That Is Beginning to Take Form in Guatemala

As a result of over 170 years of generally irresponsible governance and political violence in Guatemala, a new revolution is reportedly taking shape with the objective of achieving John Holloway's notion of taking control of government through the democratic electoral process. Additionally, sources who wish to remain anonymous report that this proposed "citizen's revolution" is intended to challenge the ruling political establishment as early as 2012.⁴⁹ The vehicle for attaining these ambitious objectives is a popular political party, the Frente Amplio (Broad Front, or Popular Front), which comprises elements of Guatemala's political "left-of-center." The intent of the leadership, reportedly, is to mobilize a popular political party (the Frente Amplio), not just for a few "true believers" but also including debourgeoised Christians, socialists, trade unionists, intellectuals, students, peasants, and members of the middle class to "march together with a majority of the population."⁵⁰ Abraham Guillen and John Holloway would argue, "Then, and only then, can a [nonviolent revolutionary political force] harness the energy of the masses and defeat the corrupted and incompetent ruling classes in democratic elections."⁵¹ This is the reported intent of the party leadership.⁵² Intent, however, does not always lead to desired objectives.

Because of the generally secretive nature of the effort to challenge the ruling establishment with the creation of a left-of-center popular political party, there is a lack of certainty regarding the activities of the leadership. Accordingly, we cannot be sure what they may or may not be doing, and why. We do know, however, what the leftist leadership *should* be doing, and why. As we watch the development of the Popular Front, that knowledge provides a reliable guide to understanding how well it is proceeding toward its objective of building a citizen's party that might successfully challenge the Guatemalan political establishment in 2012.⁵³

Operational and Tactical Requirements. Operationally, to achieve desired success, a Popular Front (citizen's revolution) must develop cadres to fill the expanding political and support components of the movement and to initiate recruiting and organizing efforts with the masses. The intent is to inculcate in the new leadership, and demonstrate to the general public, a change in tactics from the brutal insurgencies of the past to a kinder and gentler approach to relations with the voting population. The aim here is to help the citizenry perceive the incumbent regime as illegitimate and ineffective, regardless of who might lead it. Additionally and conversely, the aim is to develop public opinion, through persuasion and limited coercion, that will consider the Frente Amplio to be the legitimate and effective alternative to the incumbent political structure. That will be key to this effort.[54]

More specifically, at the operational level, the new left-of-center political effort will require the formation, nurturing, and maturation of numerous ancillary national organizations, the most important of which include a united central trade union organization, a united youth federation, and an allied labor party. The purpose of these organizations will be to continue to raise the level of direct popular support against "indigenous and *yanqui* imperialism." These organizations will also provide leadership experience and human skill that will be necessary when the time comes to establish a direct democratic government of the people and to instill a socialist mode of productions and distribution.[55] At the tactical level, the Frente Amplio will operate in small groups all over the country with the intent of supplementing operational efforts with organizations having grassroots political-psychological popular support. These operations will be carefully planned and executed with the ultimate objective of lessening the established regime's credibility and legitimacy.[56]

All these activities take time and patience to develop properly. Lenin argued that the party leadership must take all the time necessary to develop the proper foundations to achieve the ultimate objective of taking control of a targeted government. No shortcut will work.[57]

Success in the Political Conflict Environment. The left-of-center leadership will likely find the tasks of developing a united front political organization slow, difficult, and painful. In contrast to past strategies, they will have to appreciate the importance of moral as well as de facto and de jure legitimacy. They will have to understand that the vanguard of the proletariat cannot bring about a successful revolution without the active support of the people they purport to lead. They will have to rethink their ideas of tolerance, cooperation, equality, and compromise as they try to build a political coalition that could, in fact, threaten to topple the incumbent regime in democratic elections. They will also have to organize a campaign of information gathering and develop public support for a new people-oriented agenda based on this information, rather than the usual socialist rhetoric. These are some of the basic analytical commonalities that have proved to generate success in the political conflict environment over the years and throughout the world.[58]

As an example, it took the Uruguayan Tupamaros over thirty years to accomplish these tasks successfully—and they did not have to confront drug cartels, gangs, and hidden powers in the process.[59] It is, thus, hard to credit the Guatemalan Frente Amplio with possible success in the not-too-distant future. Nevertheless, a "citizen's revolution" could become a disruptive force in the political process and should not be dismissed too lightly.

The Resultant Internal Security Situation in Guatemala

Authorities have no consistent or reliable data on the TCO-gang phenomenon in Guatemala. Nevertheless, the phenomenon is acknowledged to be large and complex. Additionally, the situation differs in different parts of the country. In the north (in the Peten region and along the Mexican and Belizan borders), the area is under the virtual control of the drug barons and their mercenary enforcer gangs. Whatever state presence that exists is irrelevant. Even farther north, Central American Maras controlled the drug-trafficking land routes through Mexico to the United States in the not-too-distant past. Now the Maras are apparently being replaced or co-opted by the Zetas and their Kaibile allies (former members of the Guatemalan elite special forces). Finally, though a formidable gang presence is known to exist throughout the entire country, it is different in Guatemala City than in the south of the country. The point here is that many of the private security providers in Guatemala City are, allegedly, illicit gangs. The point in the south of the country turns on the "spillover" of illicit drug-gang activity into other Central American countries. Given the weaknesses of the Guatemalan and other Central American political, economic, social, and security institutions, criminality has considerable opportunity to prosper. Conversely, political and socioeconomic development can only diminish.[60]

The convoluted array of Central American Maras, Mexican Zetas, Guatemalan Kaibiles, and Mexican and South American drug cartels operating in Guatemala leaves an almost anarchical situation throughout the country. As gangs and TCOs compete with each other and the governments of the Central American region to maximize market share and freedom of movement and actions, we see a strategic internal security environment characterized by ambiguity, complexity, and violence. We also see the slow erosion of Guatemalan democracy and the state, coupled with the establishment of large and small criminal free enclaves in various regions, cities, and municipalities of the country.[61] At the same time, we see the beginnings of a left-of-center political movement that would attempt to implement John Holloway's basic notion of attaining power and precipitating radical political-economic-social change without resort to violence.[62]

Given the narco-gang competition, the difficulty and time required to win the support of the Guatemalan people, and the difficulty and time required to develop a successful organizational effort, the New Left leadership is unlikely to achieve its objectives as early as 2012. Even it if should, the track record of governance by the New Left is not particularly good. When they do take power, they tend to use the power of the state to stay in office rather than create a new utopia. In other instances, New Socialist and populist reformers tend to diminish the power and institutional strength of the state to the point where they become susceptible to civil war or becoming civilian or military dictatorships or narco-criminal states. In any event, there are few, if any, utopias in the world today.[63]

Again, the convoluted security situation in Guatemala reminds one of the violent political-security situation of the feudal medieval era. Violence and the fruits of violence—diffuse, arbitrary, and unprincipled political control—seem to be devolving to the criminal nonstate actors (such as the hidden powers or the TCO-gang phenomenon).[64] Guatemala's future appears to be bleak, indeed.

MAINTAINING STATE SOVEREIGNTY

Success in dealing with contemporary internal security and sovereignty issues comes as a result of a holistic effort to apply the full human and physical resources of a nation-state and its internal and international partners to achieve the individual and collective well-being that leads to sustained societal peace with justice. In the long term, however, governments such as that in Guatemala cannot depend on an international organization such as the United Nations or another country such as the United States to do these things for them. At the same time, no government can simply legislate or decree these legitimizing qualities for itself. Governments can, however, develop, sustain, and enhance these qualities by their actions over time. Legitimization and internal stability derive from popular and institutional perceptions that authority is genuine and effective, and that it uses morally correct means for reasonable and fair purposes. Establishment of morally legitimate authority and internal stability, in turn, implies a serious anticorruption campaign—and the putative power to implement it.[65]

The principal effort of a responsible government and its internal and international allies must center on obvious and comprehensive political reform. Guatemala's institutions are ill equipped to deal with popular demands in a coherent manner. Political parties have lost credibility, government agencies have little or no credibility or are simply absent, and security organizations often act more like criminal organizations than institutions that are supposed to provide legitimate personal and collective security. Well-planned and holistic reform, then, would seem to be the ideal political objective of a government aiming toward the survival of the state. If the Guatemalan people cannot see and experience meaningful change and reform, the demagogues, populists, warlords, drug lords, criminals, gangs, and other illicit organizations will likely compete for control of the failing state.[66]

Solutions to these problems require the highest level of strategic-political thought in addition to exceptional civil-military, military-police, and military-to-military diplomacy, cooperation, and coordination. Such solutions take the United States beyond unilateral training and equipping units for conducting tactical-operational-level counternarcotics, counterterrorist, and counterinsurgency operations to multilateral strategic-political-social approaches to broader professional military-police-civilian education and leader development, and organization for unity of effort.[67]

As stated explicitly and implicitly throughout this book, these concepts must also go beyond the traditional-legal definition of national security and sovereignty, and overcome external orientation and conventional military and law enforcement biases. Importantly, these concepts must also provide a point of departure from which friends and allies might advance the understanding—and the implementation—of a holistic common multilateral security agenda. The recommended basic direction of reform and institution building is well beyond the scope of this chapter, but the sooner it is authoritatively elaborated, the better.

KEY POINTS AND LESSONS

- Violent crime, impunity from prosecution, political inaction, weak institutions, and widespread corruption are generating the correlation of forces that make for an unstable security situation in Guatemala.
- These destabilizers are being created and exploited by elites who emerged out of the Peace Accords of 1996: TCOs that include Colombian and Mexican drug cartels and private armies operating in Guatemala; pandillas and Maras that are at the center of the crime epidemic afflicting the country; and the "hidden powers," which consist of corrupted businessmen and government officials who are integral elements of Guatemala's socioeconomic elite.
- Additionally, revolutionaries, including populists, twenty-first-century socialists, and former insurgents sense the development of a political-security situation that will allow them to realize their utopian dreams.
- The human and physical elements that perpetrate the destruction of security, democracy, and the state are not the ultimate threat. Rather, they are symptoms of the threat—that is, state failure.
- Violence and the fruits of violence—diffuse, arbitrary, and unprincipled political control—seem to be devolving to criminal nonstate actors rather than to democratic reformers.
- Governments, such as that of Guatemala, cannot depend on international organizations or other countries to supply the human and monetary resources necessary to achieve sustained

societal peace with justice. Moreover, no government can simply decree or legislate these legitimizing qualities. The principal effort of a responsible government and its international allies must center on obvious and comprehensive reform.
- Otherwise, competing elites will vie for control of the failing state. The results of that competition will be determined by the political actor in the field with the best ability to mobilize the center of gravity, that is, the people of a given territory.

As things stand now, criminal nonstate actors appear to have the upper hand. Guatemala's future would seem to be bleak, indeed.

In *Man and the State*, the great contemporary French political philosopher Jacques Maritain reminds us that "the highest functions of the state [are] to ensure the laws and facilitate the free development of the body politic.... [O]nly then will the State achieve its true dignity, which comes not from power and prestige, but from the exercise of justice."[68]

CHAPTER 6

Traumatic Attacks at Another Level

Cyber and Biological War

A former secretary of the U.S. Navy, Richard Danzig, reminds us that for the past several centuries, offensive warfare has aimed to destroy, degrade, or capture an opponent's troops, weapons, resources, and territory. Since the invention of gunpowder, the main means of accomplishing these objectives has been by explosive weaponry: bullets, bombs, mines, and missiles. But war is changing. The aim, more and more, is not to kill people or capture territory. Rather, the idea is to sap the ability and will of an adversary to use his superior conventional military power. The main means will be nonexplosive traumatic asymmetric war that takes us to another level of conflict. Accordingly, current and future threats arise from the use of new asymmetric means—predominantly information (cyber) and biological warfare.[1]

Other contemporary writers and thinkers, ranging from Alvin and Heidi Toffler (*War and Anti-war*, 1993) and T. X. Hammes (*The Sling and the Stone*, 2006) to Qiao Liang and Wang Xiangsui (*Unrestricted Warfare*, 1999) and Jorge Verstrynge (*La guerra asimetrica y el Islam revolucionario* [Asymmetric War and Revolutionary Islam], 2005), tell us that war will be an asymmetric "cocktail mixture" prosecuted by ways and means other than conventional arms.[2] In these terms, advancing technology threatens to turn chemical and biological agents into the so-called poor man's nuclear bomb. Biological and other types of asymmetric war are not science fiction; they are present and real. Recently, the 2009 Commission on the Prevention of Weapons of Mass Destruction

Proliferation and Terrorism warned us, "Unless the world community acts decisively and with great urgency, it is more likely than not that a weapon of mass destruction (WMD) will be used in a terrorist attack somewhere in the world by the end of 2013."[3] Clearly, regardless of the accuracy of this apocalyptic prediction, the proverbial clock is ticking.

ASYMMETRIC WAR IN CONTEXT, AND ITS HUMAN DIMENSION

First we must be clear about what we mean by contemporary asymmetric war. Then we may proceed to the human aspects of contemporary irregular asymmetric conflict, the So what? question. This is the context within which we can better understand Jorge Verstrynge's "totalization of war,"[4] the traumatic surprise suicide attacks on the World Trade Center and the Pentagon on 9/11, and the proliferation of inexpensive missiles and weapons of mass destruction (WMD). In that same context, we can better understand the expansion of contemporary conflict to include the use of nonexplosive weapons of mass destruction. Moreover, the idea of WMD under the control of terrorists who think of death as the ultimate freedom, who have no definable society, infrastructure, or command center to threaten, should send a shudder down our collective spine. In this context, a terrible asymmetry looms ahead.

Asymmetry

Contemporary asymmetric warfare is total and unrestricted—outside all rules, limitations, and conventional methods—using all conceivable ways and means to achieve one's ends.[5] This definition provides a very broad array of options ranging, for example, from flying hijacked passenger jets into the World Trade Center to using nuclear, cyber, and biological weapons to intimidate and coerce people and governments into acceding to one's demands. In these terms, the governing practices most clear today are those of the "ascension of extremes," coupled with suicide, simplicity, efficiency, and surprise in both method and ends. In these terms, it must be emphasized that there are no limitation in terms of scope or time.⁶

As a consequence, asymmetric war is one step more toward the "totalization of war." That is, total war is achieved when the number of combatants equals the number of a targeted population of fighting age, and when all their social activity has been converted for use in the war.⁷ The basic characteristics of Verstrynge's popular concept of asymmetric war include the following:

> The utilization of suicide actions;
> The achievement of both interstate and intrastate war;
> The inclusion of information war (including cyber and biological war) among the major instruments of hegemonic conflict;
> The inclusion of time among the major instruments of war;
> The elevation of divine surprises and Trojan horses to commonplace tactics in conflict;
> The "camouflaging" (integration) of military forces used as part of a combination of types of irregular asymmetric conflict within the civil population rather than having them concentrated in a predetermined geographical area; and
> The recognition of Islam—as communism once was—as a competitive world ideology.⁸

Contemporary asymmetric war, then, is much more than pitting bows and arrows against muskets or propeller-driven aircraft against jets. It is also considerably more complex. The general aim of asymmetric conflict is to accomplish what no nonstate actor or terrorist organization has previously accomplished. Far from being ingenuous, apolitical, and unique, practitioners act in accordance with a political logic that is a continuation of politics by indirect, irregular, and violent means. These practitioners do not pretend to reform an unjust order or redress perceived grievances. The intent is simply to destroy perceived regional and global enemies and replace them. This type of warfare is not a test of expertise in creating instability, conducting illegal violence, or achieving commercial, ideological, or moral satisfaction. Ultimately, it is an exercise in survival. The ultimate purpose is clear: compel radical political-economic-social systemic change.⁹

The Human Context of Contemporary Asymmetric Conflict: Ideological Romantics and Pragmatic Populists

Thanks to the revelations and events of the Hungarian Revolution of 1956, Stalinist communism was discredited. Historian Tony Judt argues that this was especially true in the territories over which the Soviet Union ruled until the European revolutions of 1989. That was the "Old Left."[10] The "New Left," as it was to call itself by 1965, sought out new directions and found them in the writings of the young Karl Marx. The metaphysical essays he wrote when barely out of his teens were steeped in the romantic dream of ultimate freedom. Additionally, the young Marx was seemingly way ahead of his time in being concerned with more-human problems: how to liberate human beings from ignorance of their true conditions and how to reverse the order of priorities in capitalist societies and place human beings at the center of their own existence; in short, how to change the world.[11] This New Left kept its distance from the Old Left but argued that the crimes of Stalinist communism were only a diversion from the crimes of capitalism.[12]

Also, during the mid-1960s, the traditional proletarian working class in Western Europe was disappearing and being replaced by a different service-oriented working population. Thus, Italian students proposed that in the new service economy, universities constituted the epicenter of knowledge production and students were now the new proletariat. Radical students declared themselves the "extra-parliamentary opposition," and New Left politics moved into the streets.[13]

The link between extraparliamentary politics and outright violence first emerged in Germany in 1968. At that time, Andreas Baader and Gudran Ensslin were arrested on suspicion of burning two department stores in Frankfurt. Two years later, Baader escaped from prison in the course of an armed raid planned and led by Utrike Meinhof. Later, she and Baader issued their "Guerrilla Manifesto" and announced the formation of a Red Army Faction (RAF) whose goal was to dismantle the German Federal Republic (West Germany) by force.[14] In Italy, the Red Brigades first came to public attention in October 1970, when they distributed leaflets describing goals that closely resembled those of the RAF. Like Baader, Meinhof, and others, the leaders of the Red Brigades were young, mostly former students, and devoted to armed struggle for its own sake.[15] Neither organization was particularly successful, and a dilemma developed.

The violence of the Old Left was not working, and the New Left was not successful either in the streets or at the ballot box. The result was that power—not a Marxian humanizing legitimacy—became the objective of both entities. Tactics, then, became more confrontational and less encumbered by judicial or political constraints.[16] Some New Socialists, however, argued that the solution to the injustice and inefficiency of capitalism was not violent urban upheaval (violence). Violence was not only undesirable and unlikely to meet its goals, it was also redundant. Genuine improvements in the condition of all classes in a society could be obtained in incremental and peaceful ways. The task of the New Socialists, as they understood it, was to "use the resources of the state to eliminate the social pathologies attendant on capitalist forms of production and distribution—to build not economic utopias, but good societies."[17]

In that connection, this part of the New Left includes populists and demagogues who tend to blame a country's problems on the poor performance of weak elected civilian leaders. Once a populist leader or a demagogue rises to power through a democratic electoral process, he or she tends to bypass, dismantle, or erode democratic institutions that restrict the concentration of power in his or her hands. If necessary, the conventional military and police power of the state is also employed to keep that leader in power. Thus, this more benign part of the New Left is also known to use totalitarian and violent methods to generate popular support to maintain or enhance its power.[18]

Through the mid-1960s to the 1980s, the West European radicals turned away from the dispiriting communist record in Eastern Europe and the idealistic few who advocated changing the world without violently taking power, and looked farther afield for inspiration and disciples. The revolutions in China, Cuba, Algeria, and Vietnam, as examples, were invested with the qualities and achievements lacking in European experience: that is, they were violent. The violence of Third World revolutions was interpreted as that of the young Marx's liberating violence. Jean-Paul Sartre explained that that violence was "man recreating himself" and argued that "to shoot down a perceived oppressor is to kill two birds with one stone, to destroy the oppressor and make the killer a free man: there remain a dead man and a liberated man."[19] Another New Left practitioner and thinker, Carlos Marighella, put it somewhat differently: "To join the armed struggle and to become today's terrorist ennobles the spirit."[20]

Common Results of Old and New Left Governance

Thus, we have a combination of Old Left (or hard Left) terrorists and guerrillas who would plunge the world back into the Dark Ages and other "hard Left" terrorists who would take us back to the glories of the seventh century. There are also other romantics in the New Left who would take us to wondrous new utopias. And other New Socialists would simply settle for personal political power. The common denominator of all the elements of this equation is a willingness to violently compel radical political change (or continuity) in targeted societies.

Accordingly, the common results of Old Left and New Left governance over the past several years tends to lead to

The erosion and eventual elimination of liberal democratic governance;
The erosion of state institutions and the resultant processes leading to state failure;
The establishment of military or civilian dictatorships;
The establishment of tribal states, criminal anarchy, or warlordism;
The creation of New Socialist populist or criminal states; or
The absorption, division, or reconfiguration of existing states into entirely different states or even nonstates.

It appears, then, that the intent of the Old Left and much of the New Left is based on the age-old notion of destroying old political-economic-social systems in order to build new and better ones.

Out of all this, another dilemma developed on both the revolutionary and counter-revolutionary sides of the equation. After seeing the devastating effects of U.S. military power in Desert Storm, the former Yugoslavia, and Afghanistan and Iraq, and noting the superior power of most nation-states around the globe, why would any wise state or nonstate competitor decide to fight a conventional insurgency or counterinsurgency? This takes us back to where we began—asymmetric warfare. Asymmetry, one way or another over the centuries, has been the only practical way to confront a stronger adversary. Thus, state and nonstate actors all around the world "are racing to build, buy, borrow, or burgle the most indiscriminate agents of mass lethality ever created—biological [and cyber] as well as nuclear."[21] But we are reminded that the use of "new" weapons of mass destruction in no way precludes the use of well-proven ways and means, including man himself. "Man is the active element [in revolution], techniques a passive one; between man and techniques, the human factor is decisive and the bearer of revolutionary values."[22]

THE TWO MOST PREVALENT TYPES OF ASYMMETRIC WAR

In the late twentieth and early twenty-first centuries, traumatic attacks have predominantly employed explosive munitions placed near buses, cars, airplanes, buildings, and troops. As a result, we tend to focus on explosives when we attempt to protect the security of airports, military bases, government buildings, means of transport, and other key facilities at home and abroad. Yet the dangers of today and the future predominantly arise from biological and information (cyber) weapons and only secondarily from chemical and radioactive materials. Attacks of this kind are less familiar but have grave potential for causing mass disruption, panic, and (in the case of biological weapons) deaths that could be counted in the hundreds of thousands.[23]

In dealing with these "new" kinds of threats, a line of separation cannot be drawn between military and civilian systems. The ability of the United States and its allies to project military power depends, both here and abroad, on utility, transport, telecommunications, and finance systems, which in turn depend on properly functioning civilian information systems and civilian employees. Thus, in addition to explosive weapons, the United States and the West will have to focus on biological and cyber weapons. In short, we will have to look to civilian systems, not just military personnel and operations, and we must consider our vulnerabilities at home as well as abroad. Additionally, we must deal with terrorist groups and individuals, as well as major powers.[24]

What Do Biological and Cyber Warfare Look Like?

The 2009 H1N1 (swine flu) pandemic, multiplied several times, provides an idea of what one might expect to see in a biological attack. A biological attack is the dissemination of bacteria, viruses, or toxins to cause debilitating or fatal illness through breathing, drinking, or absorption. Researchers have calculated that a millionth of a gram of anthrax will first sicken and then, within a week, kill anyone who inhales it. Moreover, a kilogram of anthrax has the potential to kill a million people. If an infectious agent such as plague or smallpox is used in an attack, a chain reaction can be induced, and the effects may be unbounded. Large numbers of people in panic, flight, and illness can quickly overwhelm regular systems of care, transportation, and communication, and society begins to unravel.[25]

The cyber-attack on Estonia in 2007 was a concerted "denial of service attack" on government, media, banks, and energy web servers. Similar attacks during Russia's short war with Georgia the next year looked more ominous, because they appeared to be coordinated with the advance of Russian military columns. Government and media websites went down, and telephone communications were jammed. In all, Georgia's ability to respond with any kind of effectiveness was reduced to near zero. Those cyber-attacks might be considered previews of the future—again, multiplied several times.[26] Richard Clarke, a former White House staffer in charge of counterterrorism and cyber security, envisages a catastrophic breakdown within fifteen minutes. He asserts that computer bugs will bring down military e-mail systems; oil pipelines and refineries will explode; air-traffic control systems will collapse; freight and metro trains will derail; financial data will be scrambled; the electrical grid will go down in the Eastern United States; and orbiting satellites will spin out of control. Security begins to break down as food becomes scarce and money runs out. The effects are much like a nuclear attack—except people will not be dead, they will be alive and well and demanding security and services, or providing them in their own ways.[27]

The Similarity of Information (Cyber) War to Biological War

It is striking how analogous information attacks are to their biological counterparts. For example, we use similar terminology when we describe a computer "virus." A single computer virus, like its biological equivalent, can have widespread and proliferating effects. Whether embedded in software in advance or disseminated near the time of use, a computer virus can destroy or distort data in the information and communication systems upon which civilian and military life depends. A single computer can launch an information attack. Unlike the limited and relatively ineffective anthrax letters delivered to the Capitol building in Washington, D.C., in 2001, an ordinary crop sprayer can generate a fatal anthrax cloud more than eighty miles long. A single leased airplane dispersing a biological agent can kill more people than died in any month of World War II. Moreover, the effects of these attacks can reoccur over substantial periods after delivery.[28]

Accordingly, biological and information attacks share more than a dozen characteristics that can make future security problems very different from those to which we have become accustomed. As examples, these traumatic attacks will not depend on, or be defeated by, armies or physical barriers. They do not require large, visible methods of production. Potent biological weapons can be made in a room and held in a vat. The forces of cyberspace can be marshaled on a desk and stored on a disk. The skills and assets required to wage this kind of war are very like those associated with legitimate civilian activities in the pharmaceutical and computer industries and are rather easily and inexpensively obtained. Once prepared, these weapons will not require missiles, shells, or other very visible, technically demanding, or expensive methods of delivery.[29]

Thus, because large financial resources, massive power, and elaborate delivery systems are not required, major nations do not have a monopoly on cyber and biological war. Second- and third-tier states, nonstate actors, and even individuals (hackers, criminals, or terrorists) can threaten their enemies with biological and information war. Furthermore, the characteristics of low visibility, delay, and natural occurrence can be exploited to leave uncertainty as to whether a military attack has occurred and—if it did—who conducted it. This makes a proper response more than difficult, covering cyber and biological war "in a thick, menacing blanket of uncertainty."[30]

Consequences of Biological and Cyber Warfare

Biological and cyber war do not discriminate between civilians and soldiers, between men and women, or between children and old folks. Nevertheless, such warfare is primarily aimed at civilian populations, and its consequences should be determined not by body counts or territory occupied but by the uncertainty, panic, and physical and social effects that ensue and the resultant molding of perceptions in the minds of the targeted public. In turn, these perceptions will likely determine victory or defeat. Danzig argues that traumatic biological and cyber-attacks are the thin end of the wedge by which public opinion can be leveraged—the hook on which perceptions can be hung. In the information age, information and telecommunications are primary weapons, and the public diplomacy exercised in the handling of the consequences of events may be more important than controlling the events themselves.[31] This takes us to the problem of what to do about such threats.

DEFENSE, DISSUASION, DETERRENCE, DISRUPTION, AND CONSEQUENCE MANAGEMENT

Efforts to defend against traumatic cyber and biological attacks demand more than the application of traditional approaches to conventional warfare. Above all, they demand challenging shifts in mind-sets and ways and means of dealing with enemies or suspected enemies. In addition to defense, there is a need to tailor strategies of dissuasion, deterrence, disruption, and consequence management to the challenges of traumatic attacks. None of the recommendations made below will completely circumvent crime, espionage, sabotage, or wars. But the implementation of those recommendations could make the world just a bit safer.

Defense

Defense against traumatic biological and/or cyber-attacks can begin by

    Rapid development and deployment of detector technology;
    Investment in antibiotic and vaccine research;
    Stockpiling of medicines and vaccines;
    Enhanced civilian and military awareness and training;
    Development of doctrine about how to preempt and, when, necessary, respond to a biological attack; and
    Improved intelligence.

Defense against information (cyber) warfare similarly demands innovative preparation. The main objective should be to prevent intrusions and alterations of data that can misdirect missiles, aircraft, ships, and spare parts and distort financial, utility, telecommunication, and other systems upon which we all depend. A deeper perception of these vulnerabilities should lead to greater investments in intelligence, research, and product development for computer and communications security. A perception of these vulnerabilities should also be integrated in readiness systems, the training of civilian and military personnel, and the exercise of information protection.[32]

Dissuasion and Deterrence

These activities are complex and highly dependent on mind-set in civilian and military leadership. Dissuasion seeks to avert the development of national and nonstate actor competitors, whereas deterrence seeks to limit the actions of a known competitor. These activities, then, are not necessarily military—although this is important. They are not necessarily negative or directly coercive—although that, too, is important. Dissuasion and deterrence are much broader than that. They can be direct and/or indirect political, diplomatic, socioeconomic, psychological-moral, and/or military-coercive activities. In their various forms and combinations of forms, these activities are attempts to influence how and what an enemy or potential enemy thinks and does, that is, dissuasion and deterrence attempt to create a state of mind that either discourages one thing or encourages something else.[33] Motives and culture thus become critical and political-military communication—and preventive diplomacy—become vital parts of the equation. In short, culturally effective ways and means must be found to convince traditional and nontraditional political actors that it is not in their interest, whatever that may be, to develop or continue negative behavior.

Disruption

Disruption may be a more useful strategy than deterrence when confronting terrorist groups and some resolute second- and third-tier states. Deterrence threatens reaction, whereas disruption is proactive. It intrudes upon would-be attackers with preemptive strikes, inspections, arrests, or such pressure of detection and restriction on freedom of movement as to preclude intended strikes. Western society, however, is uncomfortable with disruption, which threatens civil liberties, risks alienating public opinion, creates martyrs (through heavy-handedness), and cannot provide beforehand any assurance of success. It is, however, a recognized and helpful tool against terrorism. As a consequence, imaginative strategies of disruption that are closely controlled by civil authority and compliant with our own and international law must be developed.[34]

Consequence Management

In addition to defense, deterrence, dissuasion, and disruption, another approach is needed to deal with asymmetric traumatic attacks: consequence management. Despite a government's or an alliance's best efforts, successful traumatic attacks are likely to occur. Consequence management is intended to limit the effects of attacks. In the information (cyber war) context, this requires designing systems that are redundant and compartmentalized so that in the event of a successful attack, failure is "graceful" rather than catastrophic; and designing data systems that are camouflaged to confuse intruders, tagged and encoded to detect manipulation, and encrypted to minimize the benefits of intrusion. In the biological war context, consequence management requires investments in public health systems and standby medical capabilities put into place so that therapeutic regimes can be initiated before symptoms become pernicious. Last, in both information and biological defense, consequence management must include the creation of civilian and military information systems to help diminish panic and confusion. Of course, all of this requires investment in intelligence and research.[35]

Realities of Twenty-First-Century Conflict

Defensive approaches (which include consequence management) dealing with threats inherent in cyber and biological warfare must also carry with them a rethinking of the anachronistic distinctions between here and abroad and between military and civilians. Certainly, cyberspace has no geography. And anyone who doubts that biological agents can easily be imported into the United States or any other country need look no further than the flow of illegal drugs all around the world.

These challenges and tasks of asymmetric cyber and biological warfare are likely to prove to be the basic realities of twenty-first-century conflict. The consequences of failing to take them seriously are clear. Unless thinking, actions, and organization are reoriented to deal with these realities, the problems of global, regional, and subregional stability and security will resolve themselves—none will remain.

KEY POINTS AND LESSONS

The lessons regarding asymmetric traumatic attacks in the United States and the rest of the world are important and instructive. Perhaps the most important, however, is that it is only a matter of time before the United States will experience additional traumatic cyber and/or biological attack that will take war into a new and different level. Other lessons include the following:

- War is changing. The aim is, increasingly, not to kill people or capture territory but to sap the ability and will of an adversary to use conventional military power, no matter how superior. The main means will be nonexplosive traumatic war. Accordingly, current and future threats arise from the use of new asymmetric means—predominantly information (cyber) and biological warfare.
- Contemporary asymmetric warfare is total and unrestricted—outside all rules, limitations, and conventional methods—using all conceivable ways and means to achieve one's ends.
- Practitioners of contemporary asymmetric warfare (including Old Left terrorists and guerrillas along with New Left utopianists and New Socialists seeking personal power) act in accordance with a political logic that is a continuation of politics by indirect, irregular, and violent means. The common denominator of all the elements of this equation is a willingness to compel radical political change (or continuity) in targeted societies.
- State and nonstate actors all around the world are racing to acquire the most indiscriminate agents of mass lethality yet created.
- A biological attack is the dissemination of biological agents that cause debilitating illness or death when taken into the human body. The large numbers of people reacting to the threat could overwhelm regular systems of care, transportation, and communication, leading to an unraveling of society.
- The cyber-attacks on Estonia in 2007 and Georgia in 2008 might be considered previews of the future (as discussed in chapter 4). Some experts envisage a catastrophic breakdown within fifteen minutes, with massive numbers of people demanding security and services from the government or providing them by any means necessary.

- Because large-scale financial, personnel, and energy resources are not required, cyber and biological war are not restricted to major nations. Furthermore, such warfare can leave uncertainty as to whether an attack has even occurred, let alone permit a determination of who conducted it.
- Biological and cyber war do not discriminate but are aimed at civilian populations with the intent to ultimately compel that public and its leaders to acquiesce to the will of the perpetrator.
- Efforts to defend against traumatic cyber and biological attacks demand challenging shifts in mind-sets and ways and means of dealing with enemies or suspected enemies, adding to defense the strategies of dissuasion, deterrence, disruption, and consequence management.

Over the years, national security has been viewed largely in terms of conventional military defenses against external military threats. Given the contemporary global security environment, that is clearly too narrow a conception. The historical record demonstrates that the better a hegemonic power or government is at conducting the military aspects of conventional war near the top of the conflict ladder, the more a potential external enemy or internal enemy is inclined to move asymmetrically toward predominantly political-psychological (cyber-biological warfare) conflict. As a result, this conclusion espouses a forward-looking, proactive, and unified civil-military approach to protect against a country's asymmetric vulnerabilities and to sustain strategic advantages. Thus, according to General Sir Frank Kitson, as well as Jorge Verstrynge, instead of thinking of the various manifestations of contemporary war as being singularly military, one must regard them as steps in the ladder toward warfare as a whole (totalization).[36]

General Michael P. C. Carnes (USAF, ret.) also reminds us, "In the chaos of the 'new world disorder', the threat of devastating traumatic attacks on the United States, its interests, and its friends perpetrated by the former Soviet Union, China, and other nuclear powers retains a certain credibility. At the same time, however, the challenges for contemporary security and deterrence policy will intensify with the growing sophistication of biological and cyber war. These challenges to stability and security will be gravely complicated by 'non-traditional' threats and menaces emanating from rogue states, sub-state and trans-national terrorists, insurgents, illegal drug traffickers, organized criminals, warlords, militant fundamentalists, ethnic cleansers, and 1,000 other 'snakes' with a cause—and the will to conduct asymmetrical warfare."[37]

CHAPTER 7

The Road Ahead

This final chapter takes us along a proverbial road that is not easy, for there are curves and bumps and perhaps detours in the contemporary conflict environment. Certainly, many challenges will appear along the way.

The primary intent of this proverbial ride down a relatively unknown road is, simply, to explore some important aspects of contemporary and future asymmetric conflict. Such a reconnaissance of the road ahead provides a beginning point from which decision makers, policy makers, and opinion leaders might generate successes in unconventional conflicts and turn those successes into strategic victories.

THE NEW LANDSCAPE: REDEFINING SOME LONG-STANDING CONCEPTS

"War as cognitively known to most noncombatants, war as a battle in a field between men and machinery, war as a massive deciding event in a dispute in international affairs: such war no longer exists."[1] The author of this statement, General Rupert Smith of the United Kingdom, has the experience and understanding to explain further: "The old paradigm was that of interstate industrial war. The new one is the paradigm of war amongst peoples."[2] This new paradigm involves strategic confrontation among a range of combatants, not all of which are armies.

As an example, President Hugo Chavez of Venezuela and one of his mentors, Jorge Verstrynge Rojas, assert that this new paradigm of war among peoples has virtually unlimited possibilities for an asymmetric super insurgency against the United States and the West in the twenty-first century. Thus, Chavez and Verstrynge are providing political leaders—populists and neopopulists, New Socialists and disillusioned revolutionaries, and submerged *nomenklaturas* worldwide—with a relatively orthodox and sophisticated Leninist model for the conduct and implementation of what Chavez calls a modern "peoples' war" and what Verstrynge calls, simply, the "revalidation of guerrilla war."[3]

Yet when we think about the possibilities of conflict, we tend to invent for ourselves a comfortable U.S.-centric vision—a situation with battlefields that are well understood, with an enemy who looks and acts more or less as we do, and with a situation in which the fighting is done by the military. We must recognize, however, that in protecting our interests and confronting and influencing an adversary today, the situation has changed. We can see that change in several different ways, ranging from the identity of the enemy to the very nature of conflict.

Ambiguity

First, the traditional distinctions between crime, terrorism, subversion, insurgency, militia, mercenary and gang activity, and warfare are blurred. Sometimes, on lengthy and close examination, the common denominator of all those various actors is found to be the desire to control, destroy, and/or replace a targeted nation-state's political-economic-social system. Sometimes, each of these various players is only making an attempt to gain commercial (economic) advantage. Nevertheless, commercial self-enrichment normally requires some level of political influence or control and is an implicit political agenda.

Underlying these ambiguities is the fact that most of these activities tend to be *intrastate* affairs (that is, not an issue between sovereign states) that international law and convention is only beginning to address. Such intrastate conflict pits one part or several parts of one society against another. Thus, there are virtually no rules. In these wars, there is normally no formal declaration or termination of conflict, no easily identifiable enemy military formations to attack and destroy, no specific territory to take and hold, no single credible government or political actor with which to deal, no legal niceties such as mutually recognized national borders and Geneva Conventions to help control a situation, no guarantee that any agreement between or among contending authorities will be honored, and no commonly accepted rules of engagement to guide the leadership of any given state or nonstate actor.

Additionally and importantly, there is no territory that cannot be bypassed or used; no national boundaries or laws that cannot be ignored or used; no method or means that cannot be disregarded or used; no battlefield (dimension of conflict) that cannot be ignored or used; and no nation, transnational or nonstate actor, or international organization that cannot be ignored or used in some combination. This is why Qiao Liang and Wang Xiangsui call this kind of ambiguous conflict "unrestricted war."[4]

New Enemies, A New Center of Gravity, and a New Definition of "Victory"

The legal-traditional concept of threats to national security and sovereignty is based on the assumption that war is fought between geographically distinct nation-state adversaries, by means of well-equipped and easily identified military forces. Traditionally, then, the enemy is a nation-state that violates national borders and threatens the major institutions, natural resources, and external interests of another state. In these terms, the primary centers of gravity (the hub of all power on which all depends) are recognizable enemy military forces, coupled with the nation-state's industrial-technical capability to support military operations.

Experience gained from hundreds of small, uncomfortable insurgency (revolutionary) wars that have taken place over the past half-century (with more than one hundred ongoing today, globally) teaches us differently. At base, the enemy has now become any state or nonstate actor that plans and implements the multidimensional kinds of indirect and direct, nonmilitary and military, and nonlethal and lethal internal and external activities that threaten a given society's general well-being and exploit the root causes of internal instability. The primary and specific effort that ultimately breaks up and defeats an adversary's political-economic-social system and forces radical change is the multidimensional erosion of people's morale and political will. The better one protagonist is at conducting the persuasive-coercive effort, the more effective that protagonist will be relative to the opposition. Accordingly, the center of gravity has become public opinion and political decision-making leadership.

The basic reality of this new center of gravity is that information and the media (propaganda) is the primary currency by which "modern war amongst the people" is run. In the final analysis, the central idea in contemporary conflict is to influence and then control people and their values (the human terrain rather than the geographical terrain). This psychological-political effort also helps to define victory and defeat. As an example, people know it when they see and experience violent radical political change. On the other side of the same proverbial coin, people know it when nonstate actors have ceased coercive destabilization and political-psychological control activities.[5]

Power

In these public opinion and political leadership terms, the enemy becomes illicit "violence" and the physical, psychological, and human causes of that violence. Thus, the purposes of power have changed. Power is not simply "hard" (kinetic) power directed at a traditional enemy military formation or industrial complex. Power is multilayered, combining kinetic and nonkinetic political, psychological, moral, informational, economic, societal, military, police, and civil-bureaucratic activities that can be brought to bear appropriately on the causes as well as the perpetrators of violence.

This may be accomplished by those individuals familiar with Sun Tzu's "indirect approach"—with brain power, with an understanding of diverse cultures, with an appreciation of the power of dreams, and with a mental flexibility that goes well beyond traditional forms. The principal tools in this situation include intelligence operations, public diplomacy at home and abroad, information and propaganda operations, cultural manipulation measures to influence and/or control public opinion and decision-making leadership, and foreign alliances and partnerships. All this requires a mix of direct and indirect state and nonstate, military and nonmilitary, and/or lethal and nonlethal actions.

As a consequence, Qaio and Wang stress that warfare is no longer an exclusive "imperial garden" where professional soldiers alone can mingle. Nonprofessional warriors (hackers, financiers, media experts, software engineers, and so forth) and hegemonic nonstate organizations are posing a greater and greater threat to sovereign nations. From now on, soldiers will no longer have a monopoly on power. Consequently, the new civilian warriors must be included in the strategic architecture for contemporary war.[6]

Purpose and Motive Have Changed

One can no longer realistically expect to destroy or capture an enemy military formation. Enemies now conceal themselves among the population in small groups and maintain no fixed address. Thus, the nontraditional, contemporary purpose of becoming involved in a conflict is to establish conditions for achieving political-psychological rather than military objectives. Irregular enemies now seek to establish conditions that drain and exhaust their stronger opponents. In striving to establish these destabilizing conditions, opponents' tactical-level objectives center on attaining the widest freedom of movement and action. Operational-level objectives would include the achievement of short- and mid-term policy goals and to establish acceptance, credibility, and de facto legitimacy within the international community. In turn, freedom of movement and action takes us back to where we started. That is, the strategic political motive is to impose one's will on another.[7]

As examples, Al Qaeda and the Muslim Brotherhood represent a militant, revolutionary, and energetic commitment to a long-term approach to the imposition and renewal of an extremist interpretation of Islamic governance, social purpose, and tradition. Peru's Sendero Luminoso, likewise, is committed to a purposeful, long-term program for gaining control of the state and its society. Last, the strategic priority of the Central American Mara Salvatrucha ("Maras"), Mexican Zetas, and other less-known transnational criminal organizations (such as the Jamaican posses and Haitian and Brazilian gangs) is to operate a successful business enterprise. They are not intent on completely destroying the state or its institutions and replacing it with their own. Instead, they seek to "capture" the state: they want a weak entity that is moderately capable of functioning in the global community (banking, transportation, and providing the protection of "sovereign" status against other nation-states) and will allow the freedom for these nonstate actors to operate with impunity and increase profits.[8]

In sum, whether they are considered to be "political" insurgencies, "spiritual" insurgencies, "commercial" insurgencies, "criminal" insurgencies, or anything else, these kinds of confrontations are the organized application of coercion or threatened coercion intended to control, oppose, or overthrow an existing government and to bring about radical political change over the long term. To make this definition more meaningful, Abraham Guillen warns that "[a] popular victory would exchange the existing order of [social] classes, private property, social relations, internal and external politics, and a capitalist [economic] regime for a socialist society, economy, and political system. Consequently, [any kind of insurgency or] revolutionary war is total war. . . . [It is] a struggle without clemency that exacts the highest political tension."[9]

Conflict Has Become Multidimensional, Multilateral, Multiorganizational, and Total

Conflict can now involve entire populations, their neighbors, and friends. In these terms, conflict now involves a large number of national (whole of government) and international organizations, alliances, partnerships, private voluntary organizations, nongovernmental organizations, and other associated multilateral entities. Virtually any state and nonstate actor involved in dealing politically, economically, socially, morally, or militarily with complex and ambiguous threats to national and international security and well-being can become involved. And these are just the "good guys"; the number and diversity of "bad guy" players can be as large or larger. As examples, the conflicts in Colombia, Iraq, and Afghanistan are not simple military-to-military confrontations. They are indeed multidimensional, multilateral, multiorganizational, and total efforts to play effectively in the global, regional, national, and intranational security arenas.[10]

Asymmetric, irregular, and revolutionary wars and insurgencies will have new names, different motives, and different levels of violence that will be a new part of the old problem. Nevertheless, whether they are called "teapot wars" (Leslie Gelb), "camouflaged wars" (B. H. Liddell-Hart), "unrestricted wars" (Qiao and Wang), or something else, present and future asymmetric and total wars can be total on at least three different levels—scope, social geography, and time.[11]

Because time (the long term) becomes one of the many instruments of contemporary power and statecraft, peace or any other negotiations with hegemonic nonstate opponents must be approached with great caution. Everyone knows that the United States and other Western countries will tire of a given conflict and sooner or later will retire. Despite how badly a belligerent is beaten, as long as he is the "last man standing," he is the winner. In Guillen's terms (and adopted by other "New" and "Old" Socialist thinkers and practitioners), total war (the long war) includes no place for compromise or other options short of the ultimate political objective: radical political change. As a consequence, negotiations cannot be considered an end state. Rather, negotiations are tactical and operational means for gaining time. Vladimir Ilyich Lenin was straightforward in his statement, "Concessions are a new kind of war."[12]

The Prospect of War

At the beginning of the twenty-first century, much of the world is ripe for those who wish to change history, avenge grievances, find security in new political structures, or protect or reestablish old ways. Most of all, those who want to destabilize and destroy present systems to build new and better structures are not easily discouraged. They are not looking for anything tangible. They seek the realization of a dream: the Marxian rewards of history. Thus, this century, like the last, offers the prospect of war—but in the form of new wars that are total and unrestricted, outside traditional rules, limitations, and conventional methods.[13]

CURVES, BUMPS, AND POSSIBLE DEVIATIONS ALONG THE ROAD

As the parties, purposes, methods, and means that pertain to contemporary conflict have changed, so have the battlefields. Military strategists Steven Metz and Raymond Millen have argued that four distinct, yet highly interrelated battle spaces exist in the current security arena: (1) traditional, direct *interstate* war; (2) unconventional *inter-* and *intrastate* war, which tends to involve nonstate actors such as gangs, insurgents, drug traffickers, other transnational criminal organizations, nonterritorial communities, and warlords who thrive in the ungoverned or weakly governed spaces between and within various host countries; (3) unconventional *intrastate* war, which tends to involve direct versus indirect conflict between state and nonstate actors; and (4) indirect *interstate* war, which entails aggression by one nation-state against another through proxies or surrogates.[14]

Regardless of the analytical separation of the different battlefields, all state and nonstate actors involved are engaged in one common political act—political war—to politically institutionalize the acceptance of one's will. In this fragmented, complex, and ambiguous political-psychological environment, conflict must be considered and implemented as a whole. The power to deal with these kinds of situations is no longer only high-tech military or the relatively more benign police power. Instead, that power is derived by utilizing combinations of operations that broaden the ability of a nation-state (or a hegemonic nonstate actor) to protect, maintain, or achieve its vital interests. Regardless of what form a given conflict may take, war is war (one must remember that war, regardless of the form it takes, is still the means to compel an adversary to accept one's will).[15] Any single type of conflict can be combined with others to form completely new ways and means of conducting war. There is no instrument of power that cannot be mixed and matched with others. Thus, divine surprises, Trojan horses, and other means are likely to become parts of this equation.[16] The only limitation would be one's imagination. Perceived self-interest is the only constant.[17]

Combinations

Military, political, economic, informational, cultural, and technological (in addition to land, sea, air, space, electronic, biological, and international alliance) dimensions are all individual battlefields in their own right. Additionally, each dimension or its subparts—for example, economic war may be subdivided into trade war, financial war, and sanctions war—can be combined with as many others as a protagonist's organization and resources can deal with. That combining of dimensions provides considerably greater strength (power) than one or two operating by themselves. This concept can and must be applied in terms of an adversary's political-psychological-military deterrent capabilities. The interaction among multiple dimensions of conflict gives new and greater meaning to the idea of a state or nonstate actor using all available instruments of national and international power to pursue its objectives.

As only a few examples, combinations of military, transmilitary, and nonmilitary warfare would include the following:

Guerrilla war/drug war/media war
Conventional military war/network war/financial war
Biological war/cyber war/terrorist war
Trade war/information war/intelligence war
Diplomatic war/ideological war/conventional military war

The notion of combinations within the context of Qaio and Wang's *Unrestricted War*, Verstrynge's "Revalidation of Guerrilla War," or the work of another contemporary conflict theorist cannot be considered too ambiguous, too complex, too hard to deal with, or immoral. All of that may be true, but to admit that and do nothing would admit and invite defeat. In turn, such a response would likely submit one's posterity to unconscionable consequences. As one illustration of contemporary reality, I offer a vignette: the "Hezbollah Surprise of 2006." This is an account of a relatively straightforward use of combinations against a militarily stronger opponent.[18]

The Hezbollah Surprise of 2006

The Israeli military force that invaded southern Lebanon in the summer of 2006 was considered to be a world-class entity. The force had its own high-tech "shock and awe" strategy, and it—and its civilian leaders—expected a quick and easy conventional war of attrition. The intent (objective) was to completely incapacitate the Hezbollah movement in Lebanon. Instead of a quick and easy victory, however, the Israeli military force was surprised and thwarted on at least four levels.[19]

The First Level. On this level, the Israelis discovered that success in contemporary irregular asymmetric conflict cannot be reduced to

- Buying more and better, and heavier, equipment than the enemy has;
- Fielding more troops than the enemy possesses;
- Utilizing better conventionally trained and experienced leadership than Hezbollah could possibly develop;
- Developing a far more sophisticated logistical system than Hezbollah could contemplate; and
- Utilizing superior photo, electronic, and signal intelligence technology.[20]

The Second Level. The Israelis, at this level, were surprised and frustrated by

- Hezbollah's Katyusha and Al-Fajr rockets, Zelzal missiles, and antitank missiles;
- The remarkable power of a small "death-army" with low technology and high religious motivation, and the associated operational skill and effectiveness of the Hezbollah antitank squads;

- The astonishing initiative, determination, and vision of Hezbollah unconventional leadership;
- The ability of the Hezbollah military force to move freely among the Lebanese population and easily secure the relatively meager provisions they required;
- The accuracy of knowledge (human intelligence) the Hezbollah force had at its disposal concerning Israeli formations, strengths, tactics, and predispositions; and
- The decisive power and effectiveness of the Hezbollah political-psychological (media) campaign in Lebanon and the rest of the world.[21]

The Third Level. At the third level, it was found that

- Perceived moral legitimacy of purpose and behavior was the most important strategic principle operating in the conflict, in that Hezbollah was perceived as the defender of the Lebanese people;
- Military force is still a key element in determining the final outcome of a conflict, but that force must be supplemented by other dimensions of power, and the organization, equipment, training, and education to deal with the reality of existential asymmetric warfare;
- The need to isolate enemies politically and physically from external and internal sources of support cannot be ignored, because the political risk of not doing so is greater than the risk of making an effective effort;
- Human intelligence and culturally effective political-psychological information and propaganda campaigns are vital to success; and
- Unity of effort at all levels (not just a unity of military command) is essential to success.[22]

The Next Level. At a more holistic level, the lessons from the first three levels of analysis equal a sum greater than its parts. Thus, contemporary conflict requires an understanding of "warfare as a whole," in addition to the effective use of "combinations."[23] Hezbollah leadership appears to have understood this; they essentially conducted a military/

intelligence/media/diplomatic war. Israeli leadership apparently did not; they relied primarily on a conventional military war.

Israel unleashed escalating levels of military force upon and among the civilian populations in which Hezbollah had taken shelter. This singular military effort generated a double negative. First, the Israeli armed forces, more or less indiscriminately, inflicted casualties on civilians as they attacked Hezbollah forces. That kind of action was seen as brutal, disproportionate, and unnecessary in the eyes of the Lebanese population, the local and international media, and the conventionally oriented diplomatic community. Second, because of negative popular opinion and the surprising capabilities of the Hezbollah forces the Israelis did not achieve their objective of militarily incapacitating the Hezbollah organization in Lebanon.[24]

The Israelis were fighting a conventional limited military war of attrition against an opponent they did not understand and take seriously.[25] Israeli military equipment and other technology was good but not appropriate to the task. Israeli intelligence was good but not adequate to the task. And the Israeli local and international media and diplomatic effort was good but totally outclassed. Additionally, the Israelis misunderstood or ignored the strategic environment within which the war was taking place, and the various multidimensional centers of gravity were not assessed or considered in holistic terms. As a consequence, the Israeli invasion of Lebanon and confrontation against Hezbollah in 2006 was labeled an "absolute folly."[26]

Hezbollah was seen by all parties to the conflict as the winner. Primarily, this was because Hezbollah was perceived—in local, regional, and world opinion—to be the undeserving and unequal victim of the world-class Israeli armed forces. At the same time, Hezbollah was portrayed to the world as the protector and defender of the Lebanese people and was perceived as representing Arab pride and legitimate Arab social-economic-political hopes and wishes for the future. As long as Hezbollah did not abandon the field, it would be the winner. For each Hezbollah fighter killed in the 2006 war of attrition, a minimum of ten new supporters are estimated to have emerged out of the various Muslim communities around the world. So, while the Israelis were fighting a conventional military war, Hezbollah was subtly making unconventional long-term political-psychological preparations to take indirect control of the Lebanese state.[27]

DIVINE SURPRISES AND TROJAN HORSES IN ASYMMETRIC WAR

Probably the most-read revolutionary writer in the Middle East, North Africa, and Latin America, Jorge Verstrynge, tells us that war is a reflection of social disequilibrium. Therefore, whatever creates any kind of instability is a good revolutionary method. Two methods or means have proven to be exceptionally useful in generating political, economic, and social disequilibrium—Trojan horses and surprise.[28]

Disequilibrium can be created or exacerbated by various Trojan horses, which can include nonterritorial communities (small nonstates within a state), small support centers also located within a traditional nation-state, terrorists, insurgents, guerrillas, gangs, and two or three (or more) of these entities working together in combination.[29] A general example would be Al Qaeda support groups working with their allies (franchises) in Europe and the United Kingdom.[30] More specific examples would include the devastating surprise attacks on the Twin Towers in New York City and the Pentagon in Washington, D.C., in September 2001, the shocking Madrid bombing in March 2004, the terrorist bombing in London in 2005, the ongoing riots and civil violence in France, and other terroristic violence throughout Western Europe over the past several years.[31]

No matter what method or means may be employed in asymmetric irregular conflict, Verstrynge argues that war is most efficient when attacks are unexpected (a divine surprise), and simple. Thus, a revolutionary threat will always—ideally—be a surprise and always engender uncertainty, confusion, and instability. At the same time, the surprise will engender questions regarding who was responsible and the scope of response, if response is at all possible. Those who must make those risky decisions are at a definite disadvantage, for they must cope with limited intelligence, limited time, and limited effective assets. Additionally, the enemy will likely be interspersed among an innocent population and conventionally untargetable. Those who must respond to a divine surprise must further understand that the response—if any—will take place in an environment shaped by seemingly irrational perceptions, convictions, and ideas that will shape a counter-response.[32]

A more detailed example of what a small cell (gang or Trojan horse) can do is evident in the Madrid bombing. On March 11, 2004, ten rucksacks packed with explosives were detonated in four commuter trains at Madrid's Atocha train station (see chapter 3). Within minutes, that terrorist act killed or seriously injured two thousand innocent and unsuspecting people. It seemed random and senseless but in fact was traumatic and deliberate, and it substantively changed the Spanish government and its foreign and defense policy.

This particular divine surprise reminds one that similar surprises can easily be replicated by various types of Trojan horses. Additionally, it becomes clear that

Revolutionary (insurgent) or criminal nonstate organizations can take asymmetric irregular warfare into the global security arena;
Terrorism and associated suicide missions—sometimes seemingly irrational—are very practical tactics and strategy for the weak to use against the strong;
Political-psychological innovations combined with the ruthless application of violence (and with impunity) are a viable substitute for conventional war; and
Armed nonstate actors with varying motives and modes of operation are threatening the stability and existence of governments and other symbols of power all over the world.

Moreover, it is important to note that nonstate actors' primary concerns in pursuing strategic objectives center on small, loosely organized, and hard-to-eradicate networks. The intent of such actors is to create reliable infrastructure and franchise organizations that have the capability to attack symbols of power worldwide, without fear of serious reprisal. This deliberate and slow process (using Trojan horses and surprise) is also intended to facilitate the accomplishment of intermediate and long-term political-psychological objectives, one piece of human or physical terrain at a time.

This methodology is, in fact, an assault on state sovereignty and represents a quintuple threat to the authority, legitimacy, and stability of targeted governments. Generally, these threats are intended to

- Undermine the ability of a government to perform its legitimizing functions;
- Significantly change a government's foreign, defense, and other policies;
- Isolate religious or racial communities from the rest of a host nation's society;
- Begin to replace traditional state authority with alternative governance;
- Transform socially isolated human terrain into "virtual states" within the host state, without a centralized bureaucracy and no official armed forces; and
- Conduct low-cost actions calculated to maximize damage, minimize response, and display carefully staged media events that lead to the erosion of the legitimacy and stability of a targeted state's political-economic-social system.[33]

Finally, it is important to remember that the methodology of Trojan horses and divine surprise is not the sole property of Jorge Verstrynge. Antidemocratic populists, antisystem populists, antiglobalists, New Socialists, the revolutionary Left, radical Islamists, and any other hegemonic nonstate actor are all free to study it, adapt it, and use it for their own purposes.

CHALLENGES AND TASKS FOR THE ROAD AHEAD

The primary challenge, then, is to come to terms with the fact that contemporary security, at whatever level, is at its base a holistic political-diplomatic, socioeconomic, psychological-moral, and military-police effort. The corollary is to change from a singular military approach to a multidimensional, multiorganizational, multicultural, multinational, and total paradigm. That, in turn, requires a conceptual framework and an organizational structure to promulgate long-term unified civil-military planning and implementation of a strategic multidimensional concept.

The study of the fundamental nature of conflict has always been the philosophical cornerstone for understanding conventional war.[34] It is no less relevant to irregular asymmetric war. In the past, some wars, such as the Vietnam War, tended to be unrealistically viewed as providing military solutions to military problems. In the twenty-first century, the complex realities of contemporary wars (among the peoples) must be understood as holistic processes that rely on various civilian and military agencies, alliances or partnerships, and international organizations working together in an integrated fashion to achieve common, workable, and reasonable political-strategic ends (in other words, strategic clarity).

Given today's realities, failure to prepare adequately for present and future contingencies is unconscionable. At the minimum, there are five fundamental educational and organizational imperatives needed to implement the tasks noted above and deal effectively with contemporary conflict situations.

- Civilian and military leaders at all levels must learn the fundamental nature of subversion and insurgency with particular reference to the way in which military and nonmilitary, lethal and nonlethal, and direct and indirect force can be employed to achieve political ends, and the ways in which political considerations affect the use of force. Additionally, leaders need to understand the strategic and political-psychological implications of operational and tactical actions.
- Civilian and military personnel are expected to be able to operate effectively and collegially in coalitions or multinational contingents. They must also acquire the ability to deal collegially with civilian populations and local and global media. As a consequence, efforts that enhance interagency as well as international cultural awareness, such as civilian and military

exchange programs, language training programs, and combined (multinational) exercises must be revitalized and expanded.
- Leaders must learn that an intelligence capability several steps beyond the usual is required for small internal wars. This capability involves active utilization of intelligence operations as a dominant element of both strategy and tactics. Thus, civilian leaders and military commanders at all levels must be responsible for collecting and exploiting timely intelligence. The lowest echelon where adequate intelligence assets have been generally concentrated is the division or brigade. Yet such operations in most contemporary conflicts are conducted independently by battalion and smaller units.
- Nonstate political actors in any kind of intrastate or interstate conflict are likely to have at their disposal an awesome array of conventional and unconventional weaponry. The "savage wars of peace" have placed and will continue to place military forces and civilian support contingents in harm's way. Thus, leadership development must prepare peace enforcers working in compliance with chapters 6 and 7 of the United Nations Charter to also be effective war fighters.
- Governments must restructure themselves to the extent necessary to establish the appropriate political mechanisms to achieve an effective unity of effort. The intent is to ensure that the application of the various civilian and military instruments of power directly contribute to a mutually agreed-upon political end state. Generating a more complete unity of effort will require contributions at the international and multilateral levels, as well.

The above challenges and tasks are not radical. They are only the logical extensions of basic security strategy and national and international asset management. By accepting these challenges and tasks, the United States and the West can help to replace conflict with cooperation and harvest the hope and fulfill the promise that a new multidimensional paradigm offers. These cooperative efforts may not be easy to establish; however, they should prove in the mid- to long term to be far less demanding and costly in political, economic, military, and ethical terms than to continue a "business as usual" crisis-management approach to contemporary global security.

In discussing the utopian dreams and destructive activities of hegemonic state and nonstate actors, Albert Camus admonishes us to understand that "[h]e who dedicates himself to this history [the destruction of the old in order to build something new, and possibly better] dedicates himself to nothing and, in his turn, is nothing. But, he who dedicates himself to ... the dignity of mankind, dedicates himself to the earth and reaps from it the Harvest that sows its seed and sustains the world again and again."[35]

Afterword

EDWIN G. CORR

A tall, large-framed, strongly built, handsome man with a smiling, friendly face entered the office of the U.S. Ambassador to El Salvador on June 1, 1987, as one more of the parade of persons I met to discuss the civil war ensuing there, the role of the United States in that war, and the prospects for change and success. Professor (and Colonel) Max G. Manwaring had travelled from the U.S. Southern Command in Panama to interview me in connection with his research on "the correlates of success in counterinsurgency," a project of the Small Wars Operations Research Directorate (SWORD), for which he was the lead theoretician. He was also interviewing me for what later became be a prize-winning book (written with Court Prisk), *El Salvador at War: An Oral History* (1998).[1] Little did I imagine that my conversation with this amicable and brilliantly innovative officer and gentleman would be the beginning of twenty-five years of friendship and collaboration on how best to fight and win what Max termed *uncomfortable wars*, drawing on a lecture and article by General Jack Galvin.[2]

Max's pioneering and influential work in this area began in 1984 when General Maxwell R. Thurman, the vice-chief of staff of the U.S. Army, assigned to the Strategic Studies Institute (SSI) of the U.S. Army College the study of how the United States should cope with and succeed in the increasing number of counterinsurgencies in which our country was engaging. Dr. Manwaring was selected to direct the project. His original model relied on a "Delphic technique" and was baptized

"SSI-1." It was followed by increasing the number of cases studied to sixty-nine small wars during the post–World War II period (twenty-six of the sixty-nine were discarded because they did not fit the definition of insurgency that was acceptable at that time). This expanded study was published as the "SSI-2 model" and was subsequently referred to as the SWORD model and the Manwaring Paradigm.

Max interviewed me again in El Salvador on September 24, 1987, but our collaboration began in earnest after I had become a professor of political science at the University of Oklahoma (OU). Max asked me to write the introduction and the conclusion to his *Uncomfortable Wars: Toward a New Paradigm of Low Intensity Conflict* (1991).[3] At OU I not only continued with gratitude to publish with Max (and with Professor John Fishel and his wife, Kim Fishel), I also became the general editor of the University of Oklahoma Press series on international and security affairs. The series published John and Max's superb *Uncomfortable Wars Revisited* (2006) and Max's books of this magnificent trilogy, for which I write this afterword evaluating Max's contributions to counterinsurgency. John eloquently and solidly describes the worth of the trilogy in his foreword to this book.

The SWORD model (or the Manwaring Paradigm) that Max and his cohorts created and tested provides a systemic, comprehensive, and statistically validated approach upon which to conduct successfully, at both the strategic and tactical levels, counterinsurgency through what has come to be recognized as an essential "whole of government approach." Building on his research at SSI and at SWORD, and on his *Uncomfortable Wars* and other works, the military and civilian parts of the U.S. government began increasingly to adopt his approach for counterinsurgency (COIN) doctrine and, sporadically, in practice. This was achieved through the paradigm's presentation to military officers in the U.S. armed forces university system, through manuals, intermittently in training, and, at times, through lessons lamentably learned after massive commitments of U.S. forces (ignoring COIN doctrine) in conventional war strategies. The literature developed by Max and his colleagues was designed to, and has, influenced U.S. armed forces' and civilian agencies' thinking, terminology, and practice in counterinsurgency.

Max's research and publications have also impacted academia. Upon my arrival at OU in September 1988 as a diplomat-in-residence and full professor of political science, the teaching of courses, study, research, and publishing on counterinsurgency had not yet gained recognition and respect as a university area of concern. Max's *Uncomfortable Wars,* along with what was going on in the world, helped instigate greater academic interest in the field. This was reinforced by Manwaring and Fishel's "Insurgency and Counter-Insurgency: Toward a New Analytical Approach," published in *Small Wars and Insurgencies* (Spring–Summer 1993), which presented in academic terms and concepts the Manwaring Paradigm and its social science statistical validity. Increased interest in the field and Manwaring's approach was further boosted by the burgeoning body of literature produced by Max, John Fishel, and their colleagues. In the fall of 2000, Professor Ernestie Evans published in *World Affairs* an article calling attention to and lauding this new and growing field of research and study, most of it by Manwaring and his fellow practitioner/professor collaborators.

Books and articles introduced, explored, and highlighted what are now accepted and valued concepts in the study, development, and execution of counterinsurgency. My book (co-edited with Stephen Sloan) *Low Intensity Conflict: Old Threats in a New World* (1992)[4] used case studies and the paradigm to show it is applicable worldwide, not just in Latin America, where, because of the U.S. government and SOUTHCOM's involvement during the 1980s, insurgencies had provided the impetus and initial cases of study for the paradigm's development. Max's *Gray Area Phenomena: Confronting the New World Disorder* (1993)[5] emphasized that insurgencies, drug and human trafficking, and other problems of disorder in global security arise and thrive in areas where the legitimacy and presence of government are weak and dysfunctional. Manwaring and William J. Olson's edited volume *Managing Contemporary Conflict: Pillars of Success* (1996)[6] addressed the need for appropriate, effective national security structures. Ambassador David C. Miller's chapter in this book, "Back to the Future: Structuring Foreign Policy in a Post–Cold War World," was classic in laying out the three pillars for success: (1) the development of appropriate theories of engagement; (2) the development of appropriate weapons and national security organizations for implementation of the theories; and (3) the development of an executive branch management structure to ensure implementation of the right theories of engagement using the correct mix of civilian and military assets wisely.[7]

The Manwaring group, having already demonstrated the applicability in *Gray Area Phenomena* of the Manwaring Paradigm to narcotics trafficking control, in Fishel's *Civil Military Operations in the New World* (1996) and *The Savage Wars of Peace: Toward a New Paradigm of Peace Operations* (1998)[8] effectively extended and demonstrated the SWORD model's use in these areas. Max with Fishel in the edited volume *Toward Responsibility in the New World Disorder: Challenges and Lessons of Peace Operations* (1998)[9] and Max with Anthony James Joes in the edited volume *Beyond Declaring Victory and Coming Home: The Challenge of Peace and Stability Operations* (2000)[10] furthered thought and use of the paradigm in these areas.

There were many significant chapters in these books and numerous journal articles that added importance to this body of literature. I shall mention only one, which I believe to be of lasting significance: Kimbra Fishel's "Challenging the Hegemon" (2004).[11] I believe this may be the first publication specifically to point to nongovernmental organizations as now posing a threat to the existence of the current and prevailing global nation-state system, based on the primacy of "sovereign" states that evolved from the Treaty of Westphalia in 1648. I attribute this to Kim's keen insight and her familiarity with Max's and John's approaches to irregular warfare.

The extent to which the study and research of counterinsurgency is now commonly included in university curricula is attested by the number of regularly scheduled courses on the subject and by the many articles and books authored by professors. The publication by the University of Chicago Press of *The U.S. Army/Marine Corps Counterinsurgency Field Manual*, by General David Petraeus, Lt. General James F. Amos, and Lt. Colonel John A. Nagl (2007)[12] further shows the level of interest and importance now given by elite university presses to counterinsurgency. The impact of Max and his colleagues has been great, and I remain puzzled as to why in this University of Chicago Press manual's forewords, introductions, preface, and acknowledgments, Max, Fishel, and others receive little recognition for their pioneering and missionary work.

Max's most recent books published by the University of Oklahoma Press constitute what is, to me, a refined and sophisticated summary of a tested and proven strategy for the United States' decisions and actions related to counterinsurgency policy and operations. The books encompass counterinsurgency on an abstract, general, and strategic level as well as the tactical and operational level of the manuals. *Uncomfortable Wars Revisited* (2006) and the trilogy of *Insurgency, Terrorism, and Crime: Shadows from the Past and Portents for the Future* (2008), *Gangs, Pseudo-Militaries, and Other Modern Mercenaries: New Dynamics in Uncomfortable Wars* (2010),[13] and finally *The Complexity of Modern Asymmetric Warfare* (2012) crown Max's already noteworthy achievements. They carve out for him an honored place as a creative and pragmatic contributor to national and global security. Collectively, they come close to completing his life's intellectual contributions in this field—they are, as University of Oklahoma Press editor Jay Dew calls them, "Max Manwaring 'On War.'"

Dr. Manwaring started his journey of studying counterinsurgency at the behest of General Thurman to create the SSI-1 and SSI-2 models. General Jack Galvin, while commander in chief of SOUTHCOM (and later U.S. commander of the European Theater and of NATO), importantly gave high priority and support to Max's work at SOUTHCOM while he was assigned to SWORD. As the U.S. Ambassador to El Salvador, I provided encouragement to Max to continue in his worthy endeavor, and as a professor, I increased and continued my support for him. This initial and persistent interest and backing by senior officials deeply involved in counterinsurgency, coupled with Max's position in the U.S. Army's excellent institutions for university-level education and for research, have provided him access to and exchanges with major actors in insurgencies and counterinsurgency at all levels and around the world.

Most of his coworkers and contributors to his books also occupied positions that gave them similar access and views. For example, Dr. John Fishel, his coauthor on important works, has been a military officer involved in small wars and insurgencies and a professor at both U.S. Army universities and prestigious civilian institutions of higher learning. Dr. William J. Olson, a co-editor with Max, served as the deputy assistant secretary of state for international narcotics matters, as director of the Low Intensity Conflict Organization of the assistant secretary of defense for special operations and low intensity conflict, and as a senior civilian officer of the Coalition Provisional Authority in Iraq, as well as being a professor and researcher at universities and think tanks. Most of the contributors of chapters to Max's books and authors of articles for journals using the paradigm have similar backgrounds. To mention one other, Ambassador David Miller's experience on the National Security Council provided him an essential perspective on how counterinsurgencies can be managed and mismanaged from the highest level of our government. I, myself, have had a deep and abiding interest in how to cope successfully with insurgencies and irregular wars because of my own observation of and involvement in counterinsurgency in Colombia, 1966–68; in Southeast Asia, 1972–75; in narcotics control, especially in gray areas of much of the world, as the deputy assistant secretary of state for international narcotics control, 1978–80; as the U.S. Ambassador to Peru during the Shining Path guerrillas' launch of its civil war on a national scale, 1980–81; my four years as ambassador to Bolivia, which had become a "gray nation" under a narcotics trafficking military government, 1981–85; and my ambassadorship in El Salvador, 1985–88, during the civil war there.

For a small but growing number of strategic thinkers, there are lessons for Americans in examining and comparing U.S. historical approaches in terms of time and scales-of-involvement across the spectrum of the Vietnam–El Salvador–Iraq/Afghanistan wars. Todd Greentree's *Crossroads of Intervention: Insurgency and Counterinsurgency Lessons from Central America* (2008)[14] attempts to do this. I believe that examination of these wars, coupled with and in light of the Manwaring Paradigm, will suggest to the U.S. government and the American people elementary "lessons" about America's correct responses and conduct in uncomfortable, asymmetric conflicts and wars—nearly all of them not wars of choice but supposedly undertaken to protect and advance our vital national security interests. We examine and compare U.S. experiences in these wars conscious that Max continuously calls our attention to Clausewitz's enjoinder that every war is unique and different, and the first task is to know the kind of conflict in which one is engaged.

Once that is determined, Manwaring's Paradigm provides guidelines as to how best to fight that particular war based on the paradigm's seven key critical areas of actions.

Southeast Asia, particularly Vietnam—fought within the overall context of the U.S. cold war with the USSR and initially the Peoples' Republic of China—was a focus of immense U.S. government efforts from the early 1960s through 1975. The Vietnam War was publicly divisive within the United States and provoked sometimes violent public opposition. Clausewitz's and Max's counsel that the first step is to determine what kind of war we are fighting is exemplified by Vietnam. Former Secretary of Defense Robert S. McNamara lamented upon reflection years after the war, "What went wrong was a basic misunderstanding of the threat to our own security represented by the North Vietnamese.... [W]e exaggerated the threat.... We didn't know our opposition.... We didn't understand the Chinese; we didn't understand the Vietnamese, particularly the North Vietnamese. So the first lesson is know your opponents."[15] U.S. goals encompassed the defeat of the Vietcong insurgency and the creation of an acceptable and viable Vietnamese government, as well as defeating the North Vietnamese occupation of South Vietnam in a conventional war. The United States did not understand and proportionally address these "wars."

The civil wars in the small countries of Central America, 1979–92, were also controversially considered to be part of the cold war. The United States involved itself openly in support of the government of El Salvador and "covertly," yet publicly, supported the insurgent "Contras" in Nicaragua. These "interventions" were not militarily on the scale of U.S. participation in Southeast Asia in the 1960s–70s or in Iraq and Afghanistan during the first decades of the twenty-first century, but the strategic stakes in Central America were, in the U.S. government's analysis, perhaps even more vital to America because of Central America's proximity to the United States and the area's strategic value in terms of U.S. national security.

The Afghanistan and Iraq wars of 2001 to the present began in Afghanistan with American response to the Al Qaeda attack on the United States on September 11, 2001. The U.S. focus shifted to Iraq in expenditure of U.S. lives and resources with the 2003 U.S. invasion of Saddam Hussein's Iraq—primarily on the basis of faulty "intelligence" that Saddam threatened his neighbors and the United States with nuclear weapons. These two wars were waged by a fanatical extremist Muslim minority engaged in global jihad against the West. Some extremist Muslim leaders declared their ultimate goal to be the reestablishment of the caliphate and the defeat of the United States, which some termed as the "Great Satan." In this sense, the wars posed an existential threat to the United States similar to that previously from the ideological communist Soviet Bloc. Al Qaeda and its associates differ in their nonstate character, but their lethal terrorist attacks on America and in the European and Asian states manifest the serious challenge to national and global security.

The fall of Dien Bien Phu to the North Vietnamese in 1954 marked the end of France's control of the Vietnam part of Indochina. When that happened, the United States was paying for almost 80 percent of the French effort. The United States assumed responsibility from the French for containing North Vietnam's extension of communism by supporting a series of South Vietnamese dictators who had dubious acceptance among the South Vietnamese people. When President Kennedy took office in 1961, there were around 900 American advisers in Vietnam. By the time of his assassination, there were about 16,000, and by 1964 around 23,000. During 1965, the number rose to 184,000 troops, and by mid-1969, more than a half million American military were in Vietnam.[16]

Estimates of the cost of the Vietnam War to the United States are around $580 billion, with a loss of 58,000 lives. Most authorities state that the United States lost the war, although there are a growing number of revisionists and authorities who assert that an acceptable peace had been achieved by 1972 but was given away by the United States' inability to keep its treaty commitments to the South Vietnamese. Some also argue the war drained resources from the USSR that added to its eventual collapse.

It is not yet possible to provide an account of the enormous costs to the United States of our wars in Iraq and Afghanistan, but they are immense! Troop levels in Iraq reached 150,000. The last U.S. combat troops drove out of Iraq into Kuwait on December 18, 2011, thus ending the nearly nine-year war in Iraq.

Thomas E. Ricks in *Fiasco: The American Military Adventure in Iraq* (2006) said the real war—the insurgency—began in Iraq on August 7, 2003, four months after the U.S. military thought it had prevailed in Iraq. He asserted that the U.S. approach in the months after the invasion "helped spur the insurgency and make it bigger and stronger than it might have been." Ricks said the U.S. forces were "led poorly by commanders unprepared for their mission that took away from Vietnam only the lesson that it shouldn't get involved in messy counterinsurgencies."[17] Top military and civilian decision makers at first may have ignored the "lessons indicated" (not "learned") from the successes of low-intensity conflicts of the post–World War II period in which our country was engaged, as well as the low-intensity conflict doctrine and teachings by U.S. armed forces universities, but the validity of these earlier lessons was at last recognized. Reflecting on troop levels and firepower, General Petraeus is quoted in *Tell Me How This Ends: General David Petraeus and the Search for a Way Out of Iraq* (2008) by Linda Robinson as saying, "We cannot kill our way to victory."[18]

The number of American troops killed in Iraq totaled around 4,500 with over 32,000 wounded. The dollar cost is estimated to have reached over $800 billion.[19] The strategic loss has been immense. The U.S. drawdown from Afghanistan to concentrate on the war in Iraq greatly set back progress gained in Afghanistan. In terms of the Middle East, intraregional balance of power was destroyed. The Iraq War has enabled Iran to ascend to a dominant power position in the region. The war has also been great in its destructive impact on the alliance between the United States and Turkey, which had been a vital U.S. partner in the region for a half century.

The U.S. war in Afghanistan is now the longest war in American history. The peak number of American military personnel in Afghanistan transcended 100,000 along with an additional 30,000 NATO troops. The cost of the war there to the U.S. government as of May 29, 2011, was estimated at $444 billion. The number of American and NATO lives lost exceeded 2,538. At this writing, neither victory nor a palatable withdrawal is in sight. The cost of the wars in Iraq and Afghanistan is currently more than one and a quarter trillion dollars for U.S. taxpayers.[20]

To repeat Todd Greentree, there is a sense in which El Salvador represents in strategy, financial costs, and chronological time a "crossroads" between the U.S. interventions in Vietnam and in Iraq and Afghanistan. The United States' strategy and engagement in the El Salvador case, along with several others (we will briefly examine Colombia), parallels most closely the "lessons learned" from research based on the Manwaring Paradigm. Most important, the United States was clearly successful in achieving its stated national security and political goals and did so at much less cost in money resources and in American lives.

The United States was involved importantly in support of the government of El Salvador from 1981 until the Peace Accord of 1992. From 1980 to 1992, the U.S. government provided about $4 billion in assistance to El Salvador, an average of about $333 million a year for twelve years. Of this, about $3.2 billion was economic aid and about $800 million was military assistance. Fiscal year 1987 was exceptional because of the earthquake damage assistance, and our economic help reached a total of $494.6 million. Fiscal year 1988 economic aid was about $300 million. By 1991, U.S. economic assistance was down to about $100 million and falling.[21]

The United States placed a limit of 55 U.S. armed forces "trainers" at any time stationed in El Salvador, although by including military personnel assigned to the embassy's Military Assistance Group (MLGRP) and the Defense Attachés Office (DATT), plus temporarily assigned for duty trainers from SOUTHCOM or the continental United States, there were on the average about 120 U.S. military personnel per day in El Salvador. Please note that U.S. military personnel were assigned as "trainers," not as "advisers"; their rules of engagement were *not* to participate in Salvadoran military operations in combat, and they were to fire weapons *only* for training of Salvadorans or in self-defense. Because of the roles assigned to U.S. military and civilians and the rules of engagement for them, the number of American soldiers' deaths was small, and these in self-defense under attack or in accidents.

For the war in El Salvador, the U.S. president did not issue a "finding" placing a military commander (logically it would have been CINCSOUTH) in charge of American military and civilian personnel and programs, as was the case in Iraq and in Afghanistan. The perception was that the war must be won by a legitimate government of El Salvador in charge and control of its own armed forces. U.S. ambassadors there never lost sight of U.S. political goals. There was no possibility for a repeat of the much-cited statement of Colonel Harry G. Summers in the introduction to his classic *On Strategy: A Critical Analysis of the Vietnam War* (1982): "'You know you never defeated us in the battlefield,' said the American colonel. The North Vietnamese colonel pondered this remark a moment. 'That may be so,' he replied, 'but it is irrelevant.'"[22]

Sun Tzu's adage "Know your enemy. Know yourself. One hundred battles, one hundred victories"[23] remains valid. It is true that our people in El Salvador were far better prepared in terms of language and cross-cultural skills and also possessed knowledge and understanding of the political process that McNamara lamented was deficient in Vietnam and that has been greatly lacking in Iraq and in Afghanistan. It is true as well that, partly because of this, the United States was better able to select indigenous national leaders to work with who were more sharing of U.S. goals and values and the United States was able to support them better within Salvadoran society. The United States has been incapable of finding and promoting successfully leaders of similar integrity and ability in Iraq and Afghanistan (and did not in Vietnam) as we found in Salvadoran president José Napoleón Duarte or in the minister of defense, General Carlos Eugenio Vides Casanova. However, notwithstanding the differing environments, it is clear that the U.S. approach in El Salvador proved superior to our engagements in Vietnam and in Iraq and Afghanistan in terms of financial costs, lives lost, and national interests protected and promoted. Our nation faces a great challenge in our school systems and universities in providing American military and civilian leaders and personnel with a deeper understanding of other societies and cultures and with language fluency for all parts of the world.

The U.S. strategy in El Salvador, although not based directly on the Manwaring Paradigm (since it was still not fully articulated), was conducted very closely to its tenets. The legitimacy of the host government and good governance were the highest priorities. And it was clear to Americans in charge of decisions about that war that excessive

U.S. troops on the ground there would have undermined the fledgling struggling new civilian government's legitimacy (Manwaring's most important factor) and effectiveness, and probably the American public's essential support for U.S. engagement and expenditure of resources for the war.

Colombia provides another example, particularly in U.S. support for "Plan Colombia" over the past decade, which has resulted in greatly curtailing and reducing the insurgent Armed Forces for the Colombian Revolution (FARC). Comparisons of the U.S. approach in Colombia to the U.S. campaigns in Afghanistan are enlightening. Both countries are large and rugged with remote gray areas within which guerrillas can hide and operate. Guerrillas in both countries have cross-border sanctuaries, and both obtain resources through narcotics production and trafficking. The number of guerrillas in each country at their highest level has been about 1.9 insurgents in Colombia and about 2.3 insurgents in Afghanistan for every 1,000 military-aged males in their respective countries. During the 1990s, the FARC was engaging the Colombian Armed Forces in conventional battles at the battalion level. Rule of law was minimal. Government presence and services were absent throughout much of Colombia.

At the end of the 1990s, the U.S. and Colombian governments signed the Plan Colombia Agreement. This plan limited U.S. forces in the country to 800, and the level actually present has fluctuated around 250. The cost to the United States has been around $600 million per year, or about $6 billion in total. The FARC has been significantly decreased from 18,000 to about 8,000 insurgents and is now mostly restricted to remote areas of the country. U.S. armed forces casualties are at zero, although three contracted American helicopter pilots for narcotics control activities were captured by the FARC and held as prisoners for several years. Control of U.S. armed forces personnel in Colombia is under the U.S. ambassador.

These two cases, only sketchily described, show that there are better and less costly ways for the United States to fight irregular wars than the ways we did in Vietnam and have pursued in Iraq and Afghanistan. The El Salvador and Colombia cases both conform to lessons taught and learned from the Manwaring Paradigm. In both, the United States has been more successful with far less cost in dollars, lives, and prestige.[24] These cases and comparisons validate Professor Manwaring's work in that they parallel his paradigm and brought success.

To wrap up my comments, United States national security requires readiness and the achievement of Ambassador Miller's three pillars for success for the kinds of wars the United States may be required to fight. In the evolving world arena, the United States must be prepared strategically and operationally to deal with a small number of great powers in a global balance of power (China, India, Europe, Russia, Japan, and North America, along with a second tier of emerging powers). It must also prepare for containing and responding to actions by rogue states (such as North Korea and Iran). Finally, the United States must be prepared to deal with the uncomfortable, irregular and asymmetric wars involving insurgents and nonstate actors. This latest form of wars seems likely to dominate for at least another decade. The United States must also for now and in the future (as Max illustrates in this book with his chapter on cyber warfare) expand our high-technology capabilities in new areas such as computers and robotics (drones) for military purposes and keep up to date on nuclear, chemical, and biological warfare as well as adaptations to the use of space in combat. These new "star wars" aspects of war have application at all scales of military and civilian conflicts.

Thank God that Max Manwaring has provided us an intellectual approach and paradigm to win the counterinsurgency wars that now dominate our national security horizon in the most rational, most effective, and less costly way. His series of books on this type of warfare may be coming to an end. As his general editor for the University of Oklahoma Press, I hope to entice him to write one more greatly needed book—a brief manual for America's top policy and decision makers. John Fishel has already proposed a tentative title: *The 28 Strategic Articles: A Primer on Counterinsurgency and Other Uncomfortable Wars*. Perhaps we can persuade Max to write this book with us so that our nation will have yet another of his tomes to read, ponder, and, hopefully, to use in our irregular, asymmetric, uncomfortable wars when our nation's vital interests are at stake.

NOTES

FOREWORD

1. Interviews with Peruvian military and intelligence personnel, U.S. intelligence officers, and staff officers in the Intelligence Directorate of USSOUTHCOM, ca. 1988–93.
2. Clausewitz quotations from Carl von Clausewitz, *On War*, ed. and trans. Michael Howard and Peter Paret (Princeton, N.J.: Princeton University Press, 1976), 88–89. See John T. Fishel and Max G. Manwaring, *Uncomfortable Wars Revisited* (Norman: University of Oklahoma Press, 2006), 9.

INTRODUCTION

1. See the previous books in the trilogy: Max G. Manwaring, *Insurgency, Terrorism, and Crime: Shadows from the Past and Portents for the Future* (Norman: University of Oklahoma Press, 2008); and Manwaring, *Gangs, Pseudo-Militaries, and Other Modern Mercenaries: New Dynamics in Uncomfortable Wars* (Norman: University of Oklahoma Press, 2010).
2. Rupert Smith, *The Utility of Force: The Art of War in the Modern World* (New York: Alfred A. Knopf, 2007).
3. Abraham Guillen, *Philosophy of the Urban Guerrilla: The Revolutionary Writings of Abraham Guillen*, trans. and ed. Donald C. Hodges (New York: William Morrow, 1973); John Holloway, *Cambiar el mundo sin tomar al poder: El significado de la revolucion hoy* (Buenos Aires: Universidad Autonoma de Puebla, 2002); Qiao Liang and Wang Xiangsui, *Unrestricted Warfare* (Beijing: PLA Literature and Arts Publishing House, 1999); and Jorge Verstrynge Rojas, *La guerra periferica y el Islam revolucionario: Origines, reglas, y etica de la guerra asimetrica* (Madrid: El Viejo Topo, 2005).

4. Verstrynge, *La guerra periferica*, 91. Also see Mao Tse Tung, *On Protracted War* (May 1938), available at the Marxist Internet Archive Web site, www.marxists.org; and Sun Tzu, *The Art of War*, trans. Samuel B. Griffiths (Oxford: Oxford University Press, 1971), 39–40, 77–79, 122.

5. Verstrynge, *La guerra periferica*, 42–44.

6. President Chavez used this language in a charge to the National Armed Forces (FAN) to develop doctrine for fourth-generation war. It was made before an audience gathered at the Military Academy auditorium for the "First Military Forum on Fourth Generation War and Asymmetric War" in Caracas, Venezuela, and was reported in *El Universal* on April 8, 2005. Additionally, Chavez distributed copies of Verstrynge's book to all in attendance. (The author has a copy of the special edition published specifically for the Venezuelan Army.)

7. Robert K. Yin, *Case Study Research: Design and Methods*, 2nd ed. (Thousand Oaks, Calif.: SAGE Publications, 1994), 138–39.

8. Carl von Clausewitz, *On War*, ed. and trans. Michael Howard and Peter Paret (Princeton, N.J.: Princeton University Press), 88–89.

9. Verstrynge, *La guerra periferica*, 49.

10. Steven Metz and Raymond Millen, *Future War/Future Battlespace* (Carlisle Barracks, Penn.: Strategic Studies Institute, 2003), ix, 15–17.

11. Paul E. Smith, *On Political War* (Washington, D.C.: National Defense University Press, 1989), 3.

12. General Sir Frank Kitson, *Warfare as a Whole* (London: Faber and Faber, 1987); and General Sir Rupert Smith, *The Utility of Force: The Art of War in the Modern World* (New York: Alfred A. Knopf, 2007).

CHAPTER 1. SALIENT ANTECEDENTS TO THE PRESENT ARRAY OF CONFLICTS

1. Carl von Clausewitz, *On War* [1832], ed. and trans. Michael Howard and Peter Paret (Princeton, N.J.: Princeton University Press, 1976), 88.

2. Ibid.

3. Ibid., 596.

4. V. I. Lenin, "Tasks of Russian Social-Democrats," 3–11; "Socialism and War," 183–95; "Report on War and Peace," 545; "April Thesis," 295–300; "The State and Revolution," 375–76; and "Tasks of the Youth Leagues," 661–74; all in *The Lenin Anthology*, ed. Robert C. Tucker (New York: W. W. Norton, 1975).

5. See preceding note. Also see Lenin, "On Revolutionary Violence and Terror," 425; "Symptoms of a Revolutionary Situation," 275–76; and "A Great Beginning," 478.

6. Lenin, "On Revolutionary Violence and Terror," 429.

7. Alistair Horne, *A Savage War of Peace* (New York: Viking Press, 1978).

8. Ernesto "Che" Guevara, *Guerrilla Warfare* (New York: Monthly Review Press, 1961), 15–20. Also see Lenin, "Symptoms of a Revolutionary Situation," 275–77.

9. See preceding note.

10. An interview with Juan Chacon, in Marlene Dixon and Suzanne Jonas, eds., *Revolution and Intervention in Central America* (San Francisco, Calif.: Synthesis Publications, 1983), 41, 43. Also see Joaquin Villalobos, "El estado actual de la guerra y sus perspectivas," *ECA: Estudios Centroamericanos* (March 1986): 169–204.

11. Anthony James Joes, "French Algeria: The Victory and the Crucifixion of an Army," in Joes, *From the Barrel of a Gun: Armies and Revolutions* (Washington, D.C.: Pergamon-Brassey's, 1986), 139–58.

12. Matthew Connelly, "Rethinking the Cold War and Decolonization: The Grand Strategy of the Algerian War for Independence," *International Journal of Middle Eastern Studies* 33 (2001): 221–45. Also see interviews conducted by Colonel Al Baker, U.S. Army, in Paris, February 1986, in Max G. Manwaring and John T. Fishel, "Insurgency and Counter-Insurgency: Toward a New Analytical Approach," *Small Wars and Insurgencies* (Winter 1992): 272–310.

13. See preceding note. Also see Roger Trinquier, *Modern Warfare: A French View of Counterinsurgency* (Barcelona, Spain: n.p., 1965), 31–34; and Manuel Ortega, "Antecedentes de la guerra asimetrica: La guerra revolucionaria en Argelia," annex 6 in Jorge Verstrynge Rojas, *La guerra periferica*, 165.

14. Manwaring and Fishel, "Insurgency and Counter-Insurgency." Also see Qiao Liang and Wang Xiangsui, *Unrestricted Warfare* (Beijing: PLA Literature and Arts Publishing House, 1999), 118, 153, 169.

15. Trinquier, *Modern Warfare*.

16. Tony Judt, *Post War: A History of Europe since 1945* (New York: Penguin Press, 2005), 283–87.

17. Joes, *From the Barrel*, 145.

18. This assertion is made in the *procesverbaux* of the June 1958 Tunis Conference, reported in *Harbi, Les Archives*, pp. 414–26.

19. Connelly, "Rethinking the Cold War," 225.

20. Ibid., 239.

21. Ibid., 229, 237–38. Also see Joes, *From the Barrel*, 146; and Judt, *Post War*, 286–87.

22. Jean Larteguy, *The Centurians* (New York: E. P. Dutton, 1961), 181–82.

23. General Sir Rupert Smith, *The Utility of Force: The Art of War in the Modern World* (New York: Alfred A. Knopf, 2007), 3.

24. Clausewitz, *On War*, 596.

25. Qiao and Wang, *Unrestricted Warfare*, 152–69.

26. Ibid., 143. Also see General Sir Frank Kitson, *Warfare as a Whole* (London: Faber and Faber, 1969).

27. Smith, *Utility of Force*, 415.

28. Interview with Joaquin Villalobos, commander-in-chief of the Revolutionary People's Army (ERP), conducted by Marta Harnecker, in Marlene Dixon and Susanne Jonas, eds., *Revolution and Intervention in Central America* (San Francisco, Calif.: Synthesis Publications, 1983), 69–105; interview with Rafael Menjivar, spokesman for the FDR, in ibid., 63–69. Also see Joaquin Villalobos, "El estado actual de la guerra y sus perspectivas," *ECA: Estudios Centroamericanos*, no. 449 (March 1986): 169–204.

29. Author interviews in San Salvador, July 1987, with General Jose Guillermo Garcia, former Salvadoran minister of defense, and in San Salvador, December 1986, with General Jaime Abdul Gutierrez, a member of the civil-military junta that took control of the Salvadoran government in the coup of October 15, 1979.

30. See preceding note.

31. Author interviews with Dr. Alvaro Magana, former provisional president of El Salvador, in San Salvador, December 1986, June and November 1987, and February, July, and October 1989.

32. "The Role of Unity in the Revolutionary War," an interview with Juan Chacon, former member of the Executive Committee of the FDR, in Dixon and Jonas, *Revolution and Intervention*, 40–46.

33. Author interviews with Garcia; Gutierrez; and José Napoleón Duarte, president of El Salvador, in San Salvador, November 1987. Also see José Napoleón Duarte, *Duarte: My Story* (New York: G. P. Putnam's Sons, 1986), 98–104.

34. Author interviews with Garcia and Gutierrez.

35. Author interviews with Magana.

36. Author interviews with Dr. Luigi R. Einaudi, director of the Office of Policy Planning and Coordination, Bureau of Inter-American Affairs, U.S. Department of State, in Washington, D.C., September 1987; Ambassador Deane Hinton, former U.S. ambassador to El Salvador, Washington, D.C., September 1987; and General Wallace H. Nutting, former commander-in-chief, U.S. Southern Command, Orlando, Florida, January 1987 and May 1988.

37. Author interviews with Magana.

38. Harnecker interview with Villalobos, in Dixon and Jonas, *Revolution and Intervention*; and author interviews with Garcia.

39. Interviews with Chacon and Menjivar in Dixon and Jonas, *Revolution and Intervention*.

40. Interviews with Colonel Joseph S. Stringham III, former commander, U.S. Military Group in El Salvador, conducted by Colonel Charles A. Carlton, Jr., U.S. Army, in Carlisle Barracks, Pennsylvania, May 1985.

41. Author interviews with Magana.

42. Harnecker interview with Villalobos in Dixon and Jonas, *Revolution and Intervention*. Also see Guillermo M. Ungo, "The People's Struggle," *Foreign Policy* (Fall 1983): 51–63; and author interviews with Ungo (president of the FDR) in Panama City, Panama, December 1987.

43. Author interviews with Magana; Gutierrez; General Juan Bustillo, commander of the Salvadoran Air Force, in San Salvador, January 1987; Colonel Carlos Reynaldo Lopez Nuila, former vice-minister of public security, in San Salvador, December 1986; and General Carlos Eugenio Vides Casanova, former Salvadoran minister of defense, in San Salvador, December 1987.

44. Author interviews with Magana and Gutierrez; Harnecker interview with Villalobos in Dixon and Jonas, *Revolution and Intervention*.

45. Author interviews with General Fred F. Woerner, in San Francisco, November 1986, and again in Panama, March 1989.

46. Author interviews with Lopez Nuila, Bustillo, Garcia, and Vides Casanova.

47. Carlton interviews with Stringham; author interviews with Magana and Einaudi.

48. Author interviews with Magana, Lopez Nuila, and Ungo.

49. Author interviews with Ambassador Thomas Pickering, former U.S. ambassador to El Salvador, in Tel Aviv, Israel, August 1987.

50. Author interviews with Magana, Lopez Nuila, and Ungo.

51. Chacon interview, in Dixon and Jonas, *Revolution and Intervention*, 42.

52. *Duarte: My Story*, 170.

53. The assertions in this section of the text regarding the progress of the Salvadoran conflict from 1985 to 1992 are the result of interviews with approximately forty senior Salvadoran and U.S. officials. These interviews were conducted by the author between October 1986 and December 1987 and in July and August 1994 in the United States and Panama.

54. General Nguyen Giap, *Peoples' War, Peoples' Army* (New York: Frederick A. Praeger, 1962), 36–38.

55. See author interview with Ungo.

56. See Villalobos, "El estado actual." Also see author interview with Miguel Castallanos, former insurgent commander (1975–85), September 1987, in San Salvador; interview with President Duarte, November 1987, in San Salvador; and interview with Ambassador Edwin G. Corr, June 2011, in Washington, D.C.

57. See preceding note.

58. See Max G. Manwaring, "From Defeat to Power in Four Hard Lessons," in Manwaring, *Insurgency, Terrorism, and Crime: Shadows from the Past and Portents for the Future* (Norman: University of Oklahoma Press, 2008), 157–85.

59. Sun Tzu, *The Art of War*, trans. Samuel B. Griffiths (Oxford: Oxford University Press, 1963), 77. Also see John Holloway, *Cambiar el mundo sin tomar el poder* (Buenos Aires: Universidad Autonoma de Puebla, 2002).

CHAPTER 2. NEW "KINDLER AND GENTLER" REVOLUTIONARY LESSONS FROM PERU

1. "The *Latinobarómetro* Poll: Democracy and the Downturn," *Economist*, November 15, 2008, pp. 46–47; "The *Latinobarómetro* Poll: The Democratic Routine," *Economist*, December 4, 2010, p. 51.

2. Jorge Verstrynge Rojas, *La guerra periferica y el Islam revolucionario* (Madrid, Spain: El Viejo Topo, 2005), 39–51.

3. Sun Tzu, *The Art of War*, trans. Samuel B. Griffiths (Oxford: Oxford University Press, 1971); Carl von Clausewitz, *On War* [1832], ed. and trans. Michael Howard and Peter Paret (Princeton, N.J.: Princeton University Press, 1976); and V. I. Lenin, "Report on War and Peace," in *The Lenin Anthology*, ed. Robert C. Tucker (New York: W. W. Norton, 1975), 545.

4. Verstrynge, *La guerra periferica*. Also see Lenin, "What Is to Be Done?" 38; "The Tasks of the Youth Leagues," 671–74; "Socialism and War," 188; "On Revolutionary Violence and Terror," 425; and "Report on War and Peace," 545–49; all in *The Lenin Anthology*. Also see Mao Tse Tung, *On Prolonged War* (Peking: Foreign Languages Press, 1954); Ernesto "Che" Guevara, *Guerrilla Warfare* (Lincoln: University of Nebraska Press, 1985); and Ernesto Guevara, *Che Guevara on Revolution: A Documentary Overview* (Coral Gables, Fla.: University of Miami Press, 1969).

5. Hubert Herring, *A History of Latin America*, 3rd ed. (New York: Alfred A. Knopf, 1972), 593–99.

6. Manuel González Prada, ed., *Prosa menuda* (Buenos Aires: Ediciones Imán, 1941), 156.

7. Herring, *History of Latin America*, 600–601, 607.

8. Ibid., 608–609, 611.

9. Abimael Guzmán, "El Discurso del Dr. Guzmán," in *Los partidos políticos en el Perú*, ed. Rogger Mercado U. (Lima: Ediciones Latinoamericanas, 1985), 85–90; Comité Central del Partido Comunista del Perú, *Desarrollar la guerra popular sirviendo a la revolución mundial* (Lima: Comité Central del Partido Comunista del Perú, 1986), 82–88.

10. Comité Central del Partido Comunista del Perú, *Bases de discusión* (Lima: Comité Central del Partido Comunista, 1987); "El documento oficial del Sendero," in *Interview with Chairman Gonzalo* (San Francisco, Calif.: Red Banner Editorial House, 1988).

11. Vanda Tarazana-Sevilleno, *Sendero Luminoso and the Threat of Narcoterrorism* (New York: Praeger, 1990), 111.

12. Guzmán, "El Discurso."

13. Albert Camus, *The Rebel* (New York: Vintage Books, 1956), 234, 287–88.

14. Guzmán, "El Discurso."

15. Thomas A. Marks, *Maoist Insurgency since Vietnam* (London: Frank Cass, 1966), 260.

16. Jeremy McDermott, "The Shining Path Glimmers Again," *Jane's Intelligence Review*, 2004.

17. U.S. Department of State, Office of the Coordinator of Terrorism, *Country Reports on Terrorism 2009*, ch. 6, "Terrorist Organizations" (available at http://www.state.gov/s/ct/rls/cnt/2009/130900.htm).

18. Lenin, "What Is to Be Done?" in *The Lenin Anthology*, 12–114. Also see Guzmán, "El Discurso"; and Comité Central, *Desarrollar la guerra popular*.

19. The symptoms are (1) inertia of the ruling class; (2) more than normal aggravation of popular grievances; and (3) a rise in the level of dissension of the masses. See Lenin, "Symptoms of a Revolutionary Situation," 275–77.

20. Mercado U., *Los partidos políticos*. Also see Gustavo Gorriti, "The War of the Philosopher-King," *New Republic*, June 1990, 122.

21. See notes 3 and 4, this chapter. Also see Lenin, "April Thesis," 295–300; and "The State and Revolution," 311–98; both in *The Lenin Anthology*. A good, succinct summary of Mao's stages of revolutionary war may be found in Marks, *Maoist Insurgency*, 137–38.

22. This and other quotations and assertions dealing with the sixth stage of Guzmán's strategic plan are based on author interviews conducted with high-ranking and midlevel civilian and military officers, journalists, and academics from May 1988 through July–August 2009 in Peru. Also see Simon Strong, *Shining Path: Terror and Revolution in Peru* (New York: Random House, 1992), 225–31; and Research Institute for the Study of Conflict and Terrorism, *Shining Path: A Case Study in Ideological Terrorism*, no. 260 (London: Research Institute for the Study of Conflict and Terrorism, April 1993), 1–2, 23–26.

23. Author interviews. A good discussion of the five stages of Sendero's original general plan is in David Scott Palmer, "The Sendero Rebellion in Rural Peru," in *Latin American Insurgencies*, ed. Georges Fauriol, 67–96 (Washington, D.C.: National Defense University Press, 1985). For a primary document, see Comité Central, *Bases de discusión*. Also see Gorriti, "War of the Philosopher-King."

24. See previous note. Also see Gordon H. McCormick, "The Shining Path and Peruvian Terrorism," RAND P-7297 (Santa Monica, Calif.: RAND, 1987), 11.

25. Author interviews (see note 22, this chapter).

26. Mercado U., *Los partidos políticos*.

27. Guzmán, "El Discurso"; and Comité Central, *Desarrollar la guerra popular*.

28. See note 22, this chapter.

29. Maria Elena Hidalgo, "Carlos Basombrío: No hay que magnificar pero tampoco subestimar a SL," *La República*, June 29, 2003. Also see W. Alejandro

Sanchez, "The Rebirth of Insurgency in Peru," *Small Wars and Insurgencies* (Autumn 2003): 185–98.

30. Enrique Obando, "Peru's Shining Path Survives," *Jane's* 2006; *Jane's World Insurgency and Terrorism*, May 8, 2009; and author interviews.

31. Note that the traditional cultural-geographical divide in Peru remains intact. See "Minister: Shining Path Remains a Threat in Peru's Jungles," *Latin American Herald Tribune*, August 7, 2009.

32. Douglas Farah, "The Re-emergence of the Shining Path: Latin America in a Downward Spiral," at http://douglasfarah.com/article/470/the-re-emergence-of-the-shining-path-latin-america-in-a-downward-spiral; Douglas Farah, "The Re-emergence of the Shining Path: The Criminal-Terrorist Nexus," at http://counterterrorismblog.org/12-nov-08 and at http://www.nefafoundation.org/miscellaneous/FeaturedDocs/nefafarcirnetworkdeception0908.pdf; and Joshua Partlow, "In Peru, a Rebellion Reborn," *Washington Post Foreign Service*, November 12, 2008, p. A12. Also see *Jane's World Insurgency and Terrorism*, May 8, 2009.

33. See previous note. Also see Jeremy McDermott, "The Shining Path Glimmers Again," *Jane's Intelligence Review*, January 1, 2004; "Peru's Shining Path Gaining Ground?" *Jamestown Foundation*, September 11, 2007; and author interviews.

34. See note 22, this chapter.

35. "El Documento Oficial de Sendero," in Mercado U., *Los partidos políticos*, 110.

36. See notes 3 and 4, this chapter. Also see Lenin, "Tasks of the Youth Leagues," 668; and "Capitalist Discords and Concessions Policy," 628–34; both in *The Lenin Anthology*.

37. Verstrynge, *La guerra periférica*, 42–44.

38. Author interviews. Also see notes 9 and 10, this chapter.

39. Abraham Guillen, *Philosophy of the Urban Guerrilla: The Revolutionary Writings of Abraham Guillen*, trans. and ed. Donald C. Hodges (New York: William Morrow, 1973), 231, 249, 253, 283.

40. General Vo Nguyen Giap, "The Factors of Success," in *People's War, People's Army: The Viet Cong Insurrection Manual for Underdeveloped Countries* (New York: Frederick A. Praeger, 1962), 35.

41. See notes 9 and 10, this chapter.

42. See notes 36, 37, and 40, this chapter.

43. See note 22, this chapter.

44. Max G. Manwaring and John T. Fishel, "Insurgency and Counter-Insurgency: Toward a New Analytical Model," *Small Wars and Insurgencies* (Winter 1992): 272–310.

45. *Latin American Herald Tribune*, January 28, 2010.

46. Author interviews. Also see Lenin, "Capitalist Discords and Concessions Policy," 628–34.

47. "Las Farc mantienen contactos con grupos armadas bolivarianos de Venezuela desde 2002," *El Tiempo*, February 9, 2010 (available at www.eltiempo.com/Colombia/justicia/las-farc-mantienen-contactos).

48. "En el valle de la cocaine," *Dossier Especial* 6, no. 71 (July 2011): 82–84 (available at www.defdigital.com.ar).

49. Author interviews. Also see Peter Lupsha, "Towards an Etiology of Drug Trafficking and Insurgent Relations," *International Journal of Comparative and Applied Criminal Justice* (Fall 1989): 63; and Lupsha, "The Role of Drugs and Drug Trafficking in the Invisible Wars," in *International Terrorist Operations Issues*, ed. Richard Ward and Harold Smith (Chicago: University of Chicago Press, 1987), 181.

50. Douglas W. Payne, "The Drug Super State in Latin America," *Freedom at Risk* (March–April 1989): 7–10.

51. Ibid. Also see David C. Jordan, *Drug Politics: Dirty Money and Democracies* (Norman: University of Oklahoma Press, 1999), 132–37; William J. Olson, "International Organized Crime: The Silent Threat to Sovereignty," *Fletcher Forum* (Summer/Fall 1997): 75–78; and Max G. Manwaring, "Security of the Western Hemisphere: International Terrorism and Organized Crime," *Strategic Forum* (April 1998): 2–5.

52. Verstrynge, *La guerra periferica*, 49.

53. Guillen, *Philosophy*, 233.

54. Ibid.; Verstrynge, *La guerra periferica*, 2005; and Qiao Liang and Wang Xiangsui, *Unrestricted Warfare* (Beijing: PLA Literature and Arts Publishing House, 1999).

55. Verstrynge, *La guerra periferica*, 42–44.

56. Guillen, *Philosophy*, 267.

57. See note 22, this chapter.

58. This statement is taken from a speech delivered by General Gustavo Reyes Rangel Briceno, ministro del poder popular para la defensa, en el acto de engrega de MPPPD, July 18, 2008.

59. J. Bowyer Bell, *Dragonwars* (New Brunswick, N.J.: Transaction Publishers, 1999), 417–18.

60. Lenin, "Capitalist Discords," 628.

61. Lenin, "Tasks of Russian Social-Democrats," 12–19; "Two Factors of Social–Democracy," 112–14; "Socialized War," 194–95; "The Proletarian Revolution and the Renegade Kautsky," 647; and "The Dictatorship of the Proletariat," 489–91; all in *The Lenin Anthology*. Also see Lenin, "Conducting War for Social-Democracy," 333–38; "The State and Revolution," 348–50, 371–78; and "Left-Wing Communism: An Infantile Disorder," 560-618; all in *The Lenin Anthology*.

62. Lenin, "On Revolutionary Violence and Terror," in *The Lenin Anthology*, 425.

63. See note 22, this chapter.

64. See note 22, this chapter.

65. David Passage, "Reflections on Psychological Operations: The Imperative of Engaging a Conflicted Population," in *Ideas as Weapons*, ed. G. J. David, Jr., and T. R. McKeldin III, 49–58 (Washington, D.C.: Potomac Books, 2009).

66. Ibid.

67. Ibid., 55. Also see Max G. Manwaring and Court Prisk, eds., *El Salvador at War: An Oral History* (Washington, D.C.: National Defense University Press, 1988).

68. Manwaring and Fishel, "Insurgency and Counter-Insurgency"; and Max G. Manwaring, "Reflections on the Successful Italian Counterterrorism Effort, 1968–1983," in Max G. Manwaring, *Insurgency, Terrorism, and Crime: Shadows from the Past and Portents for the Future* (Norman: University of Oklahoma Press, 2010), 186–204.

69. See previous note; also see Passage, "Reflections," 55–56.

70. See note 22, this chapter. Also see Qiao and Wang, *Unrestricted Warfare*, 133–38.

71. Camus, *The Rebel*, 234, 287–88.

72. Qiao and Wang, *Unrestricted Warfare*, 37–41.

73. Ibid., 25.

CHAPTER 3. FOUR TROJAN HORSES OF DIFFERENT COLORS

1. V. I. Lenin, "The Symptoms of a Revolutionary Situation," in *The Lenin Anthology*, ed. Robert C. Tucker (New York: W. W. Norton, 1975), 275–76.

2. Jorge Verstrynge Rojas, *La Guerra periferica y el Islam revolucionario: Origins reglas y etica de la guerra asimetrica* [Peripheral or Indirect War] (Madrid: El Viejo Topo, 2005), 69–76, 86–87. A special edition of this book was prepared for and distributed to the Army of the Bolivarian Republic of Venezuela by President Hugo Chavez in the Military Academy auditorium on the occasion of the "1st Military Forum on Fourth Generation War and Asymmetric War," April 8, 2005.

3. Ibid., 75–76.

4. Ibid. Also see Michael Scheuer, "Al Qaeda Doctrine for International Political Warfare," *Terrorism Focus* 3, no. 42 (November 1, 2006), http://jamestown.org/terrorism/news/article.php?articleid=2370189; and Scheuer, "Al Qaeda's Insurgency Doctrine: Aiming for a 'Long War,'" *Terrorism Focus* 3, no. 8 (February 28, 2006), http://jamestown.org/terrorism/news/article.php?articleid=2369915.

5. Verstrynge, *La guerra periferica*, 68–87.

6. John P. Sullivan, "Maras Morphing: Revisiting Third-Generation Gangs," *Global Crime* (August–November 2006): 493–94.

7. Max G. Manwaring, *Street Gangs: The New Urban Insurgency* (Carlisle Barracks, Penn.: Strategic Studies Institute, 2005), 2, 10–11.

8. A copy of the proceedings and charges against the twenty-nine accused can be found at http://www.elpais.es/static/especiales/2006/auto11M/elpais_auto.html?sumpag1.

9. Ibid. Also see Victoria Burnett, "Spain Arrests 16 North Africans Accused of Recruiting Militants," *New York Times*, May 29, 2007.

10. Author interviews in Madrid, Spain, July 5–8, 2006, and September 14–21, 2008. Also see "The Madrid Train Bombings," in Lorenzo Vidino, *Al Qaeda in Europe* (Amherst, N.Y.: Prometheus Books, 2006), 291–335.

11. Ibid. Also see notes 1 and 2, this chapter.

12. Jane Barrett, "Court Finds 21 Guilty of Madrid Train Bombings," Reuters, May 27, 2008. Also see Javier Jordan and Nicola Horsburg, "Mapping Jihadist Terrorism in Spain," *Studies in Conflict and Terrorism* 28, no. 3 (2005): 174–79.

13. Ibid. Also see author interviews in Madrid, Spain, July 5–8, 2006, and September 14–21, 2008.

14. Scheuer, "Al Qaeda Doctrine"; Scheuer, "Al Qaeda's Insurgency Doctrine"; and Thomas Renard, "German Intelligence Describes a 'New Quality' in Jihadi Threats," *Terrorism Focus* 5, no. 7 (February 20, 2008).

15. Author interviews with senior members of the Spanish parliament. See note 10, this chapter.

16. See previous note. Also see Scheuer, "Al Qaeda Doctrine."

17. V. I. Lenin, "The Tasks of Russian Social-Democrats," 8; "The State and Revolution," 324; and "Socialism and War," 188; all in *The Lenin Anthology*. Also see Richard Gott, *Cuba: A New History* (New Haven, Conn.: Yale University Press, 2004).

18. *The Military Balance, 2009* (London: International Institute for Strategic Studies, 2009), 75–76. Also see G. Alexander Crowther, *Security Requirements for Post Transition Cuba* (Carlisle Barracks, Penn.: Strategic Studies Institute, 2007).

19. See previous note. Also see Hal Klepak, *Cuba's Military, 1990–2005: Revolutionary Soldiers during Counter Revolutionary Times* (New York: Palgrave MacMillan, 2005).

20. See previous note.

21. Brian Latell, *After Fidel: The Inside Story of Castro's Regime and Cuba's Next Leader* (New York: Palgrave MacMillan, 2005).

22. Author interviews.

23. Ibid. Also note author interviews with General Charles E. Wilhelm, USMC (ret.), in February and June 2001. General Wilhelm had just finished

an extensive visit with Fidel and Raul Castro and other high-ranking Cuban officials.

24. See note 22, this chapter.

25. Hubert Herring, *A History of Latin America*, 3rd ed. (New York: Alfred A. Knopf, 1972), 426–42.

26. Ibid.

27. Ibid.

28. Richard L. Millett, *Searching for Stability: The U.S. Development of Constabulary Forces in Latin America and the Philippines* (Fort Leavenworth, Kans.: Combat Studies Institute Press, 2010), 47.

29. Ibid. Also see Thomas A. Bailey, *A Diplomatic History of the American People*, 9th ed. (Englewood Cliffs, N.J.: Prentice-Hall, 1974), 553–54, 676, 682, 684.

30. See previous note. Also see Joseph Napoli, "The U.S. Role in Establishing the Multinational Interim Force and the UN Stabilization Mission in Haiti (MINUSTAH)," unpublished manuscript, n.d.; and author interviews conducted in Miami, Florida, and Washington, D.C., in February 1997 and January and June 2009.

31. See previous note. Also see Max Manwaring, Donald E. Shulz, Robert Maguire, Peter Hakim, and Abigail Horn, *The Challenge of Haiti's Future* (Carlisle Barracks, Penn.: Strategic Studies Institute, 1997).

32. See previous note.

33. See note 31, this chapter. Also see Tim Padgett, "Will Criminal Gangs Take Control of Haiti's Chaos?" *Time*, January 14, 2110, www.time.com/time/specials/packages/article/D.

34. Author interviews. Also see "Crime, Violence and Development: Trends, Costs, and Policy Options in the Caribbean," Report No. 37820, Joint Report by UN Office on Drugs and Crime in the Latin American Region of the World Bank, March 2007.

35. Author interviews.

36. Ibid.

37. Ibid. Also see Manwaring et al., *Challenge of Haiti's Future*; and Action Aid, "MINUSTAH: DDR and Police, Judicial, and Correctional Reform in Haiti, Recommendations for Change" (London: Action Aid, 2006).

38. Jack Chang, "Gang Kicks Around Political Clout," *Miami Herald*, May 31, 2006; "The Mob Takes on the State," *Economist*, May 20, 2006; and Julio A. Cirino, "Marcola, el jefe del area urbana fuera de control," *Investigacion*, May 24, 2006.

39. See previous note. Also see "Sao Paulo Violence Anniversary," OSAC, May 11, 2007, at www.osac.gov/Reports/reportcfin?contentID=68155&print.

40. Abandoned by the state, *favelas* are city-states within the major cities of Brazil. They are feudal in nature and structure and are noted as such in John

Rapley, "The New Middle Ages," *Foreign Affairs* (May/June 2006): 93–103. At the same time, a *jefe da favela* may be considered roughly equivalent to a feudal baron. Also see Chang, "Gang Kicks Around"; and Cirino, "Marcola."

41. See previous note.

42. Cirino, "Marcola"; *subir al cielo* is literally translated as "go up; into heaven."

43. Ibid.

44. Police Colonel Elizeu Éclair, reported by *BBC News*, May 20, 2006, at news.bbc.co.uk/go/pr/fr/-/2/hi/Americas/4999906.stan.

45. Author interviews.

46. "Fight in the *Favelas*," *Economist*, August 4, 2007, p. 34.

47. Peter Muello, "Militias Clean Up Rio Slums," *Miami Herald*, April 30, 2007, p. 8A.

48. Jennifer M. Taw and Bruce Hoffman, *The Urbanization of Insurgency: The Potential Challenge to U.S. Army Operations* (Santa Monica, Calif.: RAND, 1994).

49. Lenin, "Tasks of Russian Social-Democrats," 4–13, including "The Revolutionary Party and Its Tactics," 8; "The State and Revolution," 324; and "Socialism and War," 188.

50. Sun Tzu, *The Art of War*, trans. Samuel B. Griffiths (Oxford: Oxford University Press, 1971), 77–78.

51. Qiao Liang and Wang Xiangsui, *Unrestricted Warfare* (Beijing: PLA Literature and Arts Publishing House, 1999), 41.

CHAPTER 4. STATE-SUPPORTED INTERNAL AND EXTERNAL PERSUASION AND COERCION

1. Alan Cullison, "Putin Rallies Youth Support; Kremlin Applies Lessons from Toppled Neighboring Governments," *Wall Street Journal*, April 12, 2005, p. A18.

2. Ibid.

3. V. I. Lenin, "The Tasks of Russian Social-Democrats," 3–11; "Two Tactics of Social-Democracy," 134–41; "Socialism and War," 194–95; "Conducting War for Social-Democracy," 333–38; and "The State and Revolution," 348–50, 371–78, all in *The Lenin Anthology*, ed. Robert C. Tucker (New York: W. W. Norton, 1975).

4. Lenin, "Socialism and War," 188. This definition is very close to that expressed by Carl von Clausewitz in *On War*, ed. and trans. Michael Howard and Peter Paret (Princeton, N.J.: Princeton University Press, 1976), 596.

5. General Rupert Smith, *The Utility of Force: The Art of War in the Modern World* (New York: Alfred A. Knopf, 2007), 375–77.

6. See note 3, this chapter.

7. Ibid.

8. Lenin, "Tasks of the Youth Leagues," 671.

9. Lenin, "Capitalist Discords and Concessions Policy," 628; and "On Revolutionary Violence and Terror," 425.

10. Lenin, "Tasks of the Youth Leagues," 661–74, in *The Lenin Anthology*.

11. Ibid., 669.

12. Lenin, "Tasks of Russian Social-Democrats."

13. Lenin, "Symptoms of a Revolutionary Situation," 275–76; "Report on War and Peace," 545; "April Thesis," 295–300; "On Revolutionary Violence and Terror," 423–32; and "A Great Beginning," 478; all in *The Lenin Anthology*. Also see note 3, this chapter.

14. Lenin, "Tasks of Russian Social-Democrats."

15. Lenin, "What Is to Be Done?" 34–38, 40, 83–87; "The State and Revolution," 324; "Report on War and Peace," 549; and "Left-Wing Communism: An Infantile Disorder," 559; all in *The Lenin Anthology*.

16. See previous note.

17. See note 15, this chapter.

18. See note 15, this chapter. Also see Lenin, "Tasks of the Youth Leagues."

19. Tony Judt, *Postwar: A History of Europe since 1945* (New York: Penguin Press, 2005), 131–39; and "The Secret Policeman's Election," *Economist*, December 8, 2007, pp. 59–60.

20. Lenin, "Tasks of Russian Social-Democrats," 3–11.

21. Lenin, "Tasks of the Youth Leagues," 671.

22. Jan Adams, "Incremental Activism in Soviet Third World Policy: The Role of the International Department of the CPSU Central Committee," *Slavic Review* (Winter 1989): 616.

23. Ibid., 615.

24. See note 1, this chapter. Also see Stephen Blank, "The Putin Succession and Implications for Russian Policy," unpublished manuscript, n.d., pp. 2–13. Also see Gleb Pavlovsky's statement, "Your task is to protect the constitutional order and physically resist attempts at a revolution," in Michael Schwirtz, "Russia's Political Youths," *Demokratizatsya* (Winter 2007): 81.

25. Alan Cullison, "Putin Rallies Youth Support," *Wall Street Journal*, April 12, 2005, p. A18; Cullison, "Russian Radicals Feel Heat," *Wall Street Journal*, September 8, 2005, p. A16; Schwirtz, "Russia's Political Youths," 73–77.

26. Cullison, "Russian Radicals Feel Heat."

27. "Putin the Terrible," *Maclain's*, September 3, 2007, pp. 32–36.

28. "We Might Bury You," from a manifesto issued by Nashi, which claims a membership of over 100,000. Translated by Leon Neyfakh, in *Harper's Magazine* (March 2001): 22.

29. Blank, "Putin Succession," 2–13, 27–28.

30. Stephen Blank, "Project 2008: Notes on the Russian Succession," *Strategic Insights* (August 2007): 1. Also see Nicole Gallina, *Law and Order in Russia:*

The Well-Arranged Police State (Zurich, Switzerland: International Relations and Security Network, Center for Security Studies, 2007).

31. Blank, "Project 2008: . . . Russian Succession," 4. Also see Owen Matthews and Ana Nemstova, "Putin's Shock Forces," *Newsweek*, May 20, 2007, p. 38; Matthews and Nemstova, "Young Russia Arises," *Newsweek International Edition*, May 28, 2007; Steven Lee Meyers, "Youth Groups Created by Kremlin Serve Putin's Cause," *New York Times*, July 8, 2007.

32. Quotations are in Matthews and Nemstova, "Putin's Shock Forces."

33. Blank, "Project 2008: . . . Russian Succession."

34. Maya Atwal, "Evaluating Nashi's Sustainability," *Europe-Asia Studies* (July 2009): 746. Also see Fred Weir, "Russia's Row with Britain Escalates," *Christian Science Monitor*, January 15, 2008, p. 4.

35. Matthews and Nemstova, "Putin's Shock Forces."

36. Matthews and Nemstova, "Young Russia Arises."

37. Steven Lee Meyers, Michael Schwirtz, and Joshua Yaffa, "Youth Groups Created by Kremlin to Serve Putin's Cause," *New York Times*, July 8, 2007, p. A1.

38. Ibid.

39. Matthews and Nemstova, "Young Russia Arises."

40. Ellen Barry, "Russians Protest Many U.S. Plots," *New York Times*, November 3, 2008, p. A11.

41. See notes 18 and 19, this chapter.

42. Allan Cullison, "Russia's Radicals Feel Heat."

43. "Old Habits, New Hypocrisy: Russia and Dissent," *Economist*, May 26, 2007.

44. "Soviet Words and Deeds," *Economist*, October 17, 2009.

45. Ibid. Also see Schwirtz, "Russian Political Youth," 80.

46. "Nashi Michki," *Russian Life* (January/February 2009): 13.

47. Qiao Liang and Wang Xiangsui, *Unrestricted Warfare* (Beijing: PLA Literature and Arts Publishing House, 1999), 123.

48. Verstrynge, *La guerra periferica*, 42–44.

49. "Cyberwar: Battle Is Joined," *Economist*, April 25, 2009, p. 20.

50. Stephen J. Blank, "Web War I: In Europe's First Information War a New Kind of War," *Comparative Strategy* 27, no. 3 (2007): 227–47.

51. Ambassador Adrian A. Besora and Jean F. Boone, "The Georgia Crisis and Continuing Democratic Erosion in Europe/Eurasia," *Foreign Policy Research Institute (FPRI)*, October 3, 2008.

52. Qiao and Wang, *Unrestricted Warfare*, 123.

53. Abraham Guillen, *Philosophy of the Urban Guerrilla: The Revolutionary Writings of Abraham Guillen*, trans. and ed. Donald C. Hodges (New York: William Morrow, 1973), 278–79.

54. Qiao and Wang, *Unrestricted Warfare*, 25.

55. "We Might Bury You," *Harper's*, March 2001.

56. Lenin, "Tasks of the Youth Leagues," 661–74.
57. Matthews and Nemstova, "Young Russia Arises."
58. Blank, "Web War I," and John Robb, "Global Guerrillas," at http://global guerillas.typepad.com/globalguerrillas.
59. Stephen Blank's conversations with Estonian authorities in Tallinn, October 2007.
60. Ibid.
61. "We Might Bury You," *Harper's*, March 2001.
62. Moscow, NTV in Russian, August 15, 2007; FBIS SOV, August 15, 2007.
63. John Markoff, "Before the Gunfire, Cyber-Attacks," *New York Times*, August 13, 2008.
64. Ibid.
65. Felix K. Chang, "Russian Military Performance in Georgia," *Foreign Policy Research Institute (FPRI)*, August 13, 2008.
66. Qiao and Wang, *Unrestricted Warfare*, 123.
67. Ibid.
68. Besora and Boone, "Georgia Crisis."
69. Owen Matthews and Anna Nemstova, "A Reputable Russia," *Newsweek International Edition*, September 1, 2008; and Andrew Kramer, "Russia Claims Its Sphere of Influence in the World," *New York Times*, September 1, 2008.
70. Albert Camus, *The Rebel* (New York: Vantage Books, 1956), 305.
71. B. H. Liddell-Hart, *Strategy* (New York: Signet, 1954), 333, 367.

CHAPTER 5. GUATEMALA AT RISK

1. Latinobarómetro, *Informe* (Santiago, Chile: Corporacion Latinobarómetro, 2008), 51, 91, 103. Also see "A Special Report on Latin America," *Economist*, September 11, 2010, pp. 1–18.
2. "Special Report on Latin America," *Economist*, September 11, 2010, 15.
3. Ibid., 5. Also see Hal Brands, *Crime, Violence, and the Crisis in Guatemala: A Case Study in the Erosion of the State* (Carlisle Barracks, Penn.: Strategic Studies Institute, 2010).
4. "Special Report on Latin America," *Economist*, September 11, 2010, p. 15.
5. V. I. Lenin, "Symptoms of a Revolutionary Situation," in *The Lenin Anthology*, ed. Robert C. Tucker (New York: W. W. Norton, 1975), 275–77; John Holloway, *Cambiar el mundo sin tomar poder* (Buenos Aires: Universidad Autonoma de Puebla, 2002).
6. Hubert Herring, *A History of Latin America*, 3rd ed. (New York: Alfred A. Knopf, 1972), 472–75. Also see Edward F. Fischer, *Guatemalan Strategic Culture* (Miami: Florida International University Latin American and Caribbean Center, 2010); Santiago Fernandez Ordonez, "Variables socio-economicas de

Guatemala in el contexto de la cultura estrategica," unpublished manuscript, June 2010; and Brands, *Crime, Violence*.

7. Herring, *History of Latin America*, 475–80.

8. The four insurgent groups, associated under an umbrella organization known as the Guatemalan National Revolutionary Unity (URNG), are the Rebel Armed Forces (FAR), the Guerrilla Army of the Poor (EGP), Organization of the People in Arms (ORPA), and the Guatemalan Labor Party (PGT). See Richard F. Nyrop, ed., *Guatemala: A Country Study* (Washington, D.C.: American University and Department of the Army, 1983), 160–62.

9. Herring, *History of Latin America*, 479, 480.

10. Nyrop, *Guatemala: A Country Study*.

11. Brands, *Crime, Violence*; Fernandez, "Variables socio-economicas"; and Fischer, *Guatemalan Strategic Culture*.

12. Brands, *Crime, Violence*, 11.

13. Ibid.; Herring, *History of Latin America*, 472–80; and Fischer, *Guatemalan Strategic Culture*.

14. Fernandez, "Variables socio-economicas"; Brands, *Crime, Violence*; and author interview with a long-time observer of Guatemalan politics, Caesar Sereseres.

15. See previous note; and Fischer, *Guatemalan Strategic Culture*.

16. Fernandez, "Variables socio-economicas," 8. Also see Miguel Giron Castillo, "Analysis de actors involucrados en acciones de oposicion a la ejecarcia de projectos energeticos y propuesta de estrategias para enfrentarles," unpublished manuscript, December 2009.

17. Fischer, *Guatemalan Strategic Culture*, 18.

18. Ibid.; Fernandez, "Variables socio-economicas," 4–8; Brands, *Crime, Violence*; and author interview with Sereseres.

19. Brands, *Crime, Violence*, 13. Also see Richard Millett and Thomas Shannon Stiles, "Peace without Security: Central America in the 21st Century," *Whitehead Journal of Diplomacy and International Relations* (Winter–Spring 2008): 31–33.

20. Brands, *Crime, Violence*, 19, 31. Also see Juan Carlos Llorca and Frank Barak, "Mexican Drug Cartels Expand Abroad," Associated Press, July 21, 2009; and Llorca and Barak, "Los cartels mexicanos tienen presencia en 17 de 22 estados de Guatemala," *El Periodico de Mexico*, December 7, 2008.

21. For a discussion of the Zetas, see Max G. Manwaring, "The Neighbors Down the Road and Across the River," in *Gangs, Pseudo-Militaries, and Other Modern Mercenaries: New Dynamics in Uncomfortable Wars* (Norman: University of Oklahoma Press, 2010), 121–47.

22. Max G. Manwaring, "Sovereignty under Siege: Gangs and Other Criminal Organizations in Central America and Mexico," in Manwaring, *Insurgency, Terrorism, and Crime: Shadows from the Past and Portents for the Future* (Norman: University of Oklahoma Press, 2008), 104–28.

23. Brands, *Crime, Violence*, 22–28.

24. Ibid., 10–11, 36. Also see Elisabeth Milkin, "Political Struggle Lays Bare the Frailty of the Guatemalan Justice System," *New York Times*, July 4, 2010, p. 4.

25. Fernandez, "Variables socio-economicas"; and Brands, *Crime, Violence*; author interview with Sereseres.

26. David C. Jordan, *Drug Politics: Dirty Money and Democracies* (Norman: University of Oklahoma Press, 1999), 193–94.

27. Brands, *Crime, Violence*; Fischer, *Guatemalan Strategic Culture*; and Fernandez, "Variables socio-economicas."

28. Stephen Krasner and Carlos Pascal, "Addressing State Failure," *Foreign Affairs* (July/August 2005): 153.

29. Marc Lacey, "Complex Defeat for Nobel Winner in Guatemala," *New York Times*, September 11, 2007, p. A-3; and Marc Lacey, "Businessman Beats Ex-General in Guatemala Voting," *New York Times*, November 5, 2007, p. A-11.

30. Author interviews; Hansen quotation from Blake Schmidt, "Ranchers and Drug Barons Threaten Rain Forest Once Ruled by the Maya," *New York Times*, July 18, 2010, p. 6.

31. See previous note.

32. Hernandez quotation from Marc Lacey, "Guatemalan Leaders under Pall in Lawyer's Killing," *New York Times*, May 22, 2009, p. A-4.

33. Ibid.

34. See note 30, this chapter.

35. Author interviews in Guatemala City, July 12–15, 2010; Lacey, "Guatemalan Leaders."

36. Milkin, "Political Struggle," p. 4.

37. Ibid.

38. Daniel C. Esty, Jack Goldstone, Ted Robert Gurr, Barbara Harff, and Pamela T. Surko, "The State Failure Project: Early Warning Research for U.S. Foreign Policy Planning," in *Preventive Measures: Building Risk Assessment and Crisis Early Warning Systems*, ed. John L. Davies and Ted Robert Gurr, 27–38 (New York: Rowman and Littlefield, 1998).

39. Author interviews and observations. Also see Jordan, *Drug Politics*; and Eduardo Pizarro and Ana Maria Bejarano, "Colombia: A Failing State?" *ReVista: Harvard Review of Latin America* (Spring 2003): 1–6.

40. See previous note. Also see Chester A. Crocker, "Engaging Failed States," *Foreign Affairs* (September/October 2003): 32–44.

41. See notes 36 and 37, this chapter; Esty et al., "State Failure Project"; and Thomas F. Homer-Dixon, *Environment, Scarcity, and Violence* (Princeton, N.J.: Princeton University Press, 1999), 133–68.

42. See previous note.

43. Francis Deng, Sadikiel Kimaro, Terrence Lyons, Donald Rothchild, and I. William Zartman, *Sovereignty as Responsibility: Conflict Management in Africa*

(Washington, D.C.: Brookings Institution, 1996). Also see Lee Feinstein and Anne-Marie Slaughter, "The Duty to Protect," *Foreign Affairs* (January/February 2004): 136–50; and Amite Etzione, "Sovereignty as Responsibility," *Orbis* (Winter 2006): 1–15.

44. Author observation; and Jordan, *Drug Politics*, 19, 142–57.

45. Jordan, *Drug Politics*.

46. John P. Sullivan, "Terrorism, Crime, and Private Armies," *Low Intensity Conflict and Law Enforcement* (Winter 2002): 239–53.

47. Jordan, *Drug Politics*, 193–94.

48. Phil Williams, *From the New Middle Ages to a New Dark Age: The Decline of the State and U.S. Strategy* (Carlisle Barracks, Penn.: Strategic Studies Institute, 2008).

49. Author interviews.

50. Abraham Guillen, *Philosophy of the Urban Guerrilla: The Revolutionary Writings of Abraham Guillen*, trans. and ed. Donald C. Hodges (New York: William Morrow, 1973), 253; and Holloway, *Cambiar el mundo*.

51. Guillen, *Philosophy*, 253.

52. Author interviews.

53. This date has been arbitrarily established and is being stated as fact. Author interviews.

54. Lenin, "Tasks of the Youth Leagues," 668, 671–79; and "Capitalist Discords and Concessions Policy," 628–34; both in *The Lenin Anthology*; and Guillen, *Philosophy*, 276, 283–84, 288.

55. Guillen, *Philosophy*, 230, 245, 249, 253, 299.

56. See Lenin, "What Is to Be Done?" 38; "Socialism and War," 188; and "Report on War and Peace," 545–49; all in *The Lenin Anthology*.

57. See Lenin citations in note 54, this chapter; also see Lenin, "Report on War and Peace."

58. See previous note.

59. Max G. Manwaring, "From Defeat to Power in Four Hard Lessons: The Uruguayan Tupamaros' Sea Change, 1962–2005," in *Insurgency, Terrorism, and Crime*, 157–85.

60. Author interviews and observations.

61. Ibid.

62. Holloway, *Cambiar el mundo*.

63. Jordan, *Drug Politics*; and Sullivan, "Terrorism, Crime."

64. See previous note. Also see Williams, *From the New Middle Ages*.

65. Deng et al., *Sovereignty as Responsibility*; Feinstein and Slaughter, "Duty to Protect"; and Etzione, "Sovereignty as Responsibility." Also see Report of the International Commission on Intervention and State Sovereignty, *The Responsibility to Protect*, dated August 15, 2001; Report of the UN Secretary General's High-Level Panel on Threats, Challenges, and Change, *A More Secure World*,

dated 2004; and Kofi Annan, "In Larger Freedom," Executive Summary, March 2005, at www.un.org.

66. Ambassador David Passage, "Reflections on Psychological Operations: The Imperative of Engaging a Conflicted Population," in *Ideas as Weapons*, ed. G. J. David, Jr., and T. R. McKeldin III, 49–58 (Washington, D.C.: Potomac Books, 2009).

67. Ambassador David C. Miller, Jr., "Back to the Future: Structuring Foreign Policy in a Post–Cold War World," in *Managing Contemporary Conflict: Pillars of Success*, ed. Max G. Manwaring and William J. Olson, 13–28 (Boulder, Colo.: Westview Press, 1996).

68. Jacques Maritain, *Man and the State* (Chicago: University of Chicago Press and Phoenix Books, 1951), 19.

CHAPTER 6. TRAUMATIC ATTACKS AT ANOTHER LEVEL

1. Richard Danzig, "Countering Traumatic Attacks," in *Deterrence in the 21st Century*, ed. Max G. Manwaring (London: Frank Cass, 2001), 98.

2. Alvin and Heidi Toffler, *War and Anti-war: Survival at the Dawn of the Twenty-first Century* (New York: Little, Brown, 1993); Colonel T. X. Hammes (USMC, ret.), *The Sling and the Stone: On War in the Twenty-first Century* (St. Paul, Minn.: Zenith Press, 2006); Qiao Liang and Wang Xiangsui, *Unrestricted Warfare* (Beijing: PLA Literature and Arts Publishing House, 1999); and Jorge Verstrynge Rojas, *La guerra asimetrica y el Islam revolucionario* (Madrid: El Viejo Topo, 2005).

3. "The Clock Is Ticking: A Progress Report on America's Preparedness to Prevent Weapons of Mass Destruction Proliferation and Terrorism," Washington, D.C., October 21, 2009.

4. Verstrynge, *La guerra periferica*, 42–43, 47–49.

5. Ibid., 28.

6. Ibid., 39–51.

7. Ibid., 27.

8. Ibid., 39–43.

9. Qiao and Wang, *Unrestricted Warfare*, 117.

10. Tony Judt, *Postwar: A History of Europe since 1945* (New York: Penguin Press, 2005), 401.

11. Ibid., 402.

12. Ibid., 420–21. The more violent fringe of the New Left was quietly financed from East Germany and the Soviet Union (through the Komsomol).

13. Ibid., 408.

14. Ibid., 470.

15. Ibid.
16. Ibid., 133–34.
17. Ibid., 363. Also see John Holloway, *Cambiar el mundo sin tomar al poder* (Buenos Aires: Universidad Autonoma de Puebla, 2002).
18. Vladimir Torres, *The Impact of "Populism" on Social, Political, and Economic Development in the Hemisphere*, FOCAL Policy Paper (Ottawa, Canada: FOCAL, 2006), 4. Also see Steve Ropp, *The Strategic Implications of the Rise of Populism in Europe and South America* (Carlisle Barracks, Penn.: Strategic Studies Institution, 2005); and Thomas Legler, *Bridging Divides, Breaking Impasses*, FOCAL Policy Paper (Ottawa, Canada: FOCAL, 2006).
19. Judt, *Postwar*, 406. The Sartre quotation is from the 1961 preface to the French edition of Frantz Fanon, *Les Damnes de la terre* (The Wretched of the Earth), Cahiers Libres Series 27–28 (Paris: François Maspero, 1961).
20. Carlos Marighella, *The Manual of the Urban Guerrilla* (Chapel Hill, N.C.: Documentary Publications, 1985), 40.
21. Toffler and Toffler, *War and Anti-war*, 192.
22. Abraham Guillen, *Philosophy of the Urban Guerrilla: The Revolutionary Writings of Abraham Guillen*, trans. and ed. Donald C. Hodges (New York: William Morrow, 1973), 302.
23. Danzig, "Countering Traumatic Attacks," 98.
24. Ibid., 104–105.
25. Ibid., 98.
26. See chapter 4, this volume.
27. Richard A. Clarke, *Cyber-War: The Next Threat to National Security and What to Do about It* (New York: Ecco, 2010).
28. Danzig, "Countering Traumatic Attacks," 99.
29. Ibid.
30. "Briefing Cyber-War," *Economist*, July 3, 2010, p. 25.
31. Danzig, "Countering Traumatic Attacks," 101.
32. Ibid., 102. Also see "The Clock Is Ticking"; and Clarke, *Cyber-War*.
33. Edwin G. Corr and Max G. Manwaring, "The Challenge of Preventive Diplomacy and Deterrence in the Global Security Environment," in Manwaring, *Deterrence in the 21st Century*, 128–29.
34. Danzig, "Countering Traumatic Attacks," 103. Also see Clarke, *Cyber-War*.
35. See previous note.
36. Frank Kitson, *Warfare as a Whole* (London: Faber and Faber, 1987). Also see Verstrynge, *La guerra periferica*.
37. General Michael P. C. Carnes, "Reopening the Deterrence Debate: Thinking about a Peaceful and Prosperous Tomorrow," in Manwaring, *Deterrence in the 21st Century*, 8.

CHAPTER 7. THE ROAD AHEAD

1. General Rupert Smith, *The Utility of Force: The Art of War in the Modern World* (New York: Alfred A. Knopf, 2007), 1.

2. Ibid., 2.

3. "War of all the people" is another translation of Hugo Chavez's words in this context. He also, interchangeably, uses "fourth-generation war." President Chavez used this language in a charge to the National Armed Forces (FAN) of Venezuela to develop doctrine for asymmetric war. It was made before an audience gathered in the Military Academy auditorium for the "First Military Forum on Fourth Generation War and Asymmetric War," in Caracas, Venezuela, and was reported in *El Universal* on April 8, 2005. At that time, it was also reported, Chavez passed out copies of Jorge Verstrynge's *La guerra periferica* to the entire audience. See Special Edition for the Army of the Bolivarian Republic of Venezuela, IDRFAN, Enlace Circular Militar (Madrid: El Viejo Topo, 2005).

4. Qiao Liang and Wang Xiangsui, *Unrestricted Warfare* (Beijing: PLA Literature and Arts Publishing House, 1999).

5. Clausewitz warned us of these factors nearly two hundred years ago. See Carl von Clausewitz, *On War* [1832], ed. and trans. Michael Howard and Peter Paret (Princeton, N.J.: Princeton University Press, 1976), 596.

6. Qiao and Wang, *Unrestricted Warfare*, 41.

7. Smith, *Utility of Force*, 275–79.

8. Max G. Manwaring, *A "New" Dynamic in the Western Hemisphere Security Environment: The Mexican Zetas and Other Private Armies* (Carlisle Barracks, Penn.: Strategic Studies Institute, 2009), 22–25.

9. Abraham Guillen, *Philosophy of the Urban Guerrilla: The Revolutionary Writings of Abraham Guillen*, trans. and ed. Donald C. Hodges (New York: William Morrow, 1973), 278–79.

10. For an elaboration of this idea, see Lt. Colonel David Last (Canadian Army), "Winning the Savage Wars of Peace," in *The Savage Wars of Peace*, ed. John T. Fishel, 211–39 (Boulder, Colo.: Westview Press, 1998).

11. Leslie H. Gelb, "Quelling the Teacup Wars," *Foreign Affairs* (November/December 1994): 2–6; B. H. Liddell-Hart, *Strategy*, 2nd ed. (New York: Praeger, 1954), 367; Qiao and Wang, *Unrestricted Warfare*.

12. V. I. Lenin, "Capitalist Discords and Concessions Policy," in *The Lenin Anthology*, ed. Robert C. Tucker (New York: W. W. Norton, 1975), 628–34. Also see Guillen, *Philosophy*, 278–79.

13. J. Bowyer Bell, *Dragonwars* (New Brunswick, N.J.: Transaction Publishers, 1999), 417–18. Also see Verstrynge, *La guerra periferica*, 21–33.

14. Steven Metz and Raymond Millen, *Future War/Future Battlespace: The Strategic Role of American Landpower* (Carlisle Barracks, Penn.: Strategic Studies Institute, 2003), 15–17.

15. Qiao and Wang remind us, "A kinder more gentle war in which brutality and bloodshed may be reduced is still war. It may alter the cruel process of war, but there is no way to change the essence of war, which is one of compulsion, and therefore it cannot alter its cruel outcome either." Qiao and Wang, *Unrestricted Warfare*, 25.

16. Verstrynge, *La guerra periferica*, 39–51, 69–87.

17. Qiao and Wang, *Unrestricted Warfare*, 105, 152, 188.

18. This vignette is also noted in Max Manwaring, *Gangs, Pseudo-Militaries, and Other Modern Mercenaries: New Dynamics in Uncomfortable Wars* (Norman: University of Oklahoma Press, 2010), 27–30.

19. "Winograd Committee Final Report," Israel Ministry of Foreign Affairs, January 30, 2008; Jeremy M. Sharp, coordinator, "Lebanon: The Israel-Hezbollah Conflict," Washington, D.C., CRS Report for Congress, September 15, 2006; and Stephen Biddle and Jeffrey A. Friedman, *The 2006 Lebanon Campaign* (Carlisle Barracks, Penn.: Strategic Studies Institute, 2008).

20. See previous note. Also see Alastair Crooke and Mark Perry, "How Hezbollah Defeated Israel, Part Two: Winning the Ground War," *Asia Times*, October 13, 2006; Crooke and Perry, "How Hezbollah Defeated Israel, Part Three: The Political War," *Asia Times*, October 14, 2006; "Hezbollah Reacts to Israel's Winograd Report," *Terrorism Focus*, Jamestown Foundation, May 8, 2007; and Matt M. Mathews, *We Were Caught Unprepared: The 2006 Hezbollah-Israeli War*, Occasional Paper 26 (Fort Leavenworth, Kans.: U.S. Army Combined Arms Center Combat Studies Institute Press, 2008).

21. See previous note.

22. "Winograd Report," commentary by Israel Ministry of Foreign Affairs, 2009. Also see Max G. Manwaring and John T. Fishel, "Insurgency and Counter-insurgency: Toward a New Analytical Approach," *Small Wars and Insurgencies* 3, no. 3 (Winter 1992): 272–310.

23. General Sir Frank Kitson, *Warfare as a Whole* (London: Faber and Faber, 1987). The best analysis of combinations, in the author's opinion, may be found in Qiao and Wang, *Unrestricted Warfare*.

24. "Winograd Report," commentary by Israel Ministry of Foreign Affairs, 2009.

25. See, as one example, Robert F. Worth and Nada Bakri, "Deal for Lebanese Factions Leaves Hezbollah Stronger," *New York Times*, May 22, 2008.

26. Ari Shavit, "A Spirit of Absolute Folly," *Haaretz*, August 15, 2006.

27. See Worth and Bakri, "Deal for Lebanese Factions."

28. Verstrynge, *La guerra periferica*, 70–71.

29. Ibid., 68–87.

30. Manwaring, *Gangs, Pseudo-Militaries*, 108–11.

31. Ibid., 91–97.

32. Verstrynge, *La guerra periferica*, 39–43.

33. Manwaring, *Gangs, Pseudo-Militaries*, 96–120 and chapter 3.
34. Kitson, *Warfare as a Whole*.
35. Camus, *The Rebel* (New York: Vintage Books, 1956), 302.

AFTERWORD

1. Max G. Manwaring and Courtney Prisk, *El Salvador at War: An Oral History* (Washington, D.C.: National Defense University Press, 1988).
2. General John R. Galvin, "Uncomfortable Wars: Toward a New Paradigm," *Parameters: The Journal of the U.S. Army War College* 16, no. 4 (December 1986): 2–8.
3. Max G. Manwaring, ed., *Uncomfortable Wars: Toward a New Paradigm of Low Intensity Conflict* (Boulder, Colo.: Westview Press, 1991).
4. Edwin G. Corr and Stephen Sloan, eds., *Low Intensity Conflict: Old Threats in a New World* (Boulder, Colo.: Westview Press, 1992).
5. Max G. Manwaring, ed., *Gray Area Phenomena: Confronting the New World Disorder* (Boulder, Colo.: Westview Press, 1993).
6. Max G. Manwaring and William J. Olson, eds., *Managing Contemporary Conflict: Pillars of Success* (Boulder, Colo.: Westview Press, 1996).
7. Ibid., 14–15.
8. John T. Fishel, *Civil Military Operations in the New World* (Westport, Conn.: Praeger, 1996); and Fishel, *The Savage Wars of Peace: Toward a New Paradigm of Peace Operations* (Boulder, Colo.: Westview Press, 1998).
9. Max G. Manwaring and John T. Fishel, eds., *Toward Responsibility in the New World Disorder: Challenges and Lessons of Peace Operations* (Portland, Ore.: Frank Cass, 1998).
10. Max G. Manwaring and Anthony James Joes, eds., *Beyond Declaring Victory and Coming Home: The Challenge of Peace and Stability Operations* (Westport, Conn.: Praeger, 2000).
11. Kimbra L. Fishel, "Challenging the Hegemon: Al Qaeda's Elevation of Asymmetric Insurgent Warfare into the Global Arena," in *Networks, Terrorism, and Global Insurgency*, ed. Robert J. Bunker, 115–28 (New York: Taylor and Francis, 2004).
12. General David H. Petraeus, Lt. General James A. Amos, and Lt. Colonel John A. Nagl, *The U.S. Army/Marine Corps Counterinsurgency Field Manual*, U.S. Army Field Manual no. 3-24 and Marine Corps Warfighting Publication no. 3-33.5. (Chicago: University of Chicago Press, 2007). The same applies to *Guiding Principles for Stabilization and Reconstruction*, United States Institute for Peace, U.S. Army Peacekeeping and Stability Operations Institute, 2009.
13. John T. Fishel and Max G. Manwaring, *Uncomfortable Wars Revisited* (Norman: University of Oklahoma Press, 2006); Manwaring, *Insurgency,*

Terrorism, and Crime: Shadows from the Past and Portents for the Future (Norman: University of Oklahoma Press, 2008); and Manwaring, *Gangs, Pseudo-Militaries, and Other Modern Mercenaries: New Dynamics in Uncomfortable Wars* (Norman: University of Oklahoma Press, 2010).

14. Todd Greentree, *Crossroads of Intervention: Insurgency and Counterinsurgency Lessons from Central America* (Westport, Conn.: Praeger Security International, 2008).

15. Tim Weiner, "Robert S. McNamara, Architect of Futile War Dies at 93," *New York Times,* July 6, 2009, http://www.nytimes.com/2009/07/07/us/07mc namara.html?pagewanted-5&th&emc=th.

16. David Levy, "The Vietnam War," *Word: Magazine of the Pioneer Library System,* (Cleveland, Okla.) (Spring 2011): 4–5.

17. Thomas E. Ricks, "Vietnam's Lessons Ignored," *Washington Post National Weekly Edition,* July 31–August 4, 2006, pp. 10, 11. See also Ricks, *Fiasco: The American Military Adventure in Iraq* (New York: Penguin Press, 2006).

18. Linda Robinson, *Tell Me How This Ends: General David Petraeus and the Search for a Way Out of Iraq* (New York: Public Affairs of Perseus Book Group, 2008). The quotation is taken from Linda Robinson, "Our Man in Baghdad," *Washington Post National Weekly Edition,* September 22–28, 2008, p. 27.

19. Deborah White, "Iraq Facts, Results, and Statistics at May 29, 2011," http://usliberals.about.com/od/homelandsecurit1/a/IraqNumbers.htm?p=1. Tim Arango and Michael S. Schmidt, "U.S. Marks End to 9-Year War, Leaving an Uncertain Iraq," *New York Times,* December 18, 2011.

20. Debby Belasco, "The Cost of Iraq, Afghanistan, and Other Global War on Terror Operations Since 9/11," Congressional Research Services, March 29, 2011; and Operation Enduring Freedom, http://icasualties.org/oef/.

21. Documents of Ambassador Edwin G. Corr.

22. Harry G. Summers, Jr., *On Strategy: A Critical Analysis of the Vietnam War* (New York: Random House, 1982), 1. This exchange was presented as transpiring in a conversation in Hanoi, April 1975.

23. Cited by Ricks on opening page of *Fiasco.*

24. Robert Haddick, "Afghanistan and Colombia: Another Tale of Two Cities," http://thegovmonitor.com/economy/afghanistan-and-colombia-an other-tale-of-two-cities-2.

Bibliography

Action Aid. *MINUSTAH: DDR and Police, Judicial, and Correctional Reform in Haiti, Recommendations for Change.* London: Action Aid, 2006.
Adams, Jan. "Incremental Activism in Soviet Third World Policy: The Role of the International Department of the CPSU Central Committee." *Slavic Review* (Winter 1989): 614–30.
Atwal, Maya. "Evaluating Nashi's Sustainability." *Europe-Asia Studies* (July 2009): 746.
Bailey, Thomas A. *A Diplomatic History of the American People.* 9th ed. Englewood Cliffs, N.J.: Prentice-Hall, 1974.
Beckett, Ian F. W., ed. *Modern Counter Insurgency.* Aldershot, England: Ashgate, 2007.
Bell, J. Bowyer. *Dragonwars: Armed Struggle and the Conventions of Modern War.* New Brunswick, N.J.: Transaction Publishers, 1999.
Biddle, Stephen, and Jeffrey A. Friedman. *The 2006 Lebanon Campaign.* Carlisle Barracks, Penn.: Strategic Studies Institute, 2008.
Blank, Stephen J. "Project 2008: Notes on the Russian Succession." *Strategic Insights* (August 2007): 1.
———. "Web War I: In Europe's First Information War a New Kind of War." *Comparative Strategy* 27, no. 3 (2007): 227–47.
Brands, Hal. *Crime, Violence, and the Crisis in Guatemala: A Case Study in the Erosion of the State.* Carlisle Barracks, Penn.: Strategic Studies Institute, 2010.
Camus, Albert. *The Rebel.* New York: Vintage Books, 1956.
Carns, Michael P. C. "Reopening the Deterrence Debate: Thinking about a Peaceful and Prosperous Tomorrow." In *Deterrence in the 21st Century,* edited by Max G. Manwaring, 7–16. London: Frank Cass, 2001.
Castañeda, Jorge. "Latin America's Left Turn." *Foreign Affairs* (May/June 2006): 28–43.

Clarke, Richard A. *Cyber-War: The Next Threat to National Security and What to Do about It.* New York: Ecco, 2010.

Clausewitz, Carl von. *On War* [1832]. Edited and translated by Michael Howard and Peter Paret. Princeton, N.J.: Princeton University Press.

Comité Central del Partido Comunista del Perú. *Bases de discusión.* Lima: Comité Central del Partido Comunista, 1987.

———. *Desarrollar la guerra popular sirviendo a la revolución mundial.* Lima: Comité Central del Partido Comunista del Perú, 1986.

Connelly, Matthew. "Rethinking the Cold War and Decolonization: The Grand Strategy of the Algerian War for Independence." *International Journal of Middle Eastern Studies* 33 (2001): 221–45.

Corr, Edwin G., and Max G. Manwaring. "The Challenge of Preventive Diplomacy and Deterrence in the Global Security Environment." In *Deterrence in the 21st Century,* edited by Max G. Manwaring, 124–30. London: Frank Cass, 2001.

Corr, Edwin G., and Stephen Sloan, eds. *Low Intensity Conflict: Old Threats in a New World.* Boulder, Colo.: Westview Press, 1992.

Crocker, Chester A. "Engaging Failed States." *Foreign Affairs* (September/October 2003): 32–44.

Crowther, G. Alexander. *Security Requirements for Post Transition Cuba.* Carlisle Barracks, Penn.: Strategic Studies Institute, 2007.

Danzig, Richard. "Countering Traumatic Attacks." In *Deterrence in the 21st Century,* edited by Max G. Manwaring, 98–105. London: Frank Cass, 2001.

Deng, Francis, et al. *Sovereignty as Responsibility.* Washington, D.C.: Brookings Institution, 1996.

Dixon, Marlene, and Susanne Jonas, eds. *Revolution and Intervention in Central America.* San Francisco, Calif.: Synthesis Publications, 1983.

Duarte, Jose Napoleon. *Duarte: My Story.* New York: G. P. Putnam's Sons, 1986.

Esty, Daniel C., Jack Goldstone, Ted Robert Gurr, Barbara Harff, and Pamela T. Surko. "The State Failure Project: Early Warning Research for U.S. Foreign Policy Planning." In *Preventive Measures: Building Risk Assessment and Crisis Early Warning Systems,* edited by John L. Davies and Ted Robert Gurr, 27–38. New York: Rowman and Littlefield, 1998.

Etzione, Amite. "Sovereignty as Responsibility." *Orbis* (Winter 2006): 1–15.

Feinstein, Lee, and Anne-Marie Slaughter. "The Duty to Protect." *Foreign Affairs* (January/February 2004): 136–50.

Fischer, Edward F. *Guatemalan Strategic Culture.* Miami: Florida International University Latin American and Caribbean Center, 2010.

Fishel, John T. *Civil Military Operations in the New World.* Westport, Conn.: Praeger, 1996.

———. *The Savage Wars of Peace: Toward a New Paradigm of Peace Operations.* Boulder, Colo.: Westview Press, 1998.

Fishel, John T., and Max G. Manwaring. *Uncomfortable Wars Revisited*. Norman: University of Oklahoma Press, 2006.
Fishel, Kimbra L. "Challenging the Hegemon: Al Qaeda's Elevation of Asymmetric Insurgent Warfare into the Global Arena." In *Networks, Terrorism, and Global Insurgency*, edited by Robert J. Bunker, 115–28. New York: Routledge, 2004.
Gallina, Nicole. *Law and Order in Russia: The Well-Arranged Police State*. Zurich, Switzerland: International Relations and Security Network, Center for Security Studies, 2007.
Galvin, John R. "Uncomfortable Wars: Toward a New Paradigm." *Parameters, the Journal of the U.S. Army War College* 16, no. 4 (December 1986): 2–8.
Gelb, Leslie H. "Quelling the Teacup Wars." *Foreign Affairs* (November/December 1994): 2–6.
Gonzalez Prada, Manuel, ed. *Prosa menuda*. Buenos Aires: Ediciones Imán, 1941.
Gorriti, Gustavo. "The War of the Philosopher-King." *New Republic*, June 18, 1990, pp. 115–22.
Gott, Richard. *Cuba, A New History*. New Haven, Conn.: Yale University Press, 2004.
Greentree, Todd. *Crossroads of Intervention: Insurgency and Counterinsurgency Lessons from Central America*. Westport, Conn.: Praeger Security International, 2008.
Guevara, Ernesto. *Che Guevara on Revolution: A Documentary Overview*. Coral Gables, Fla.: University of Miami Press, 1969.
———. *Guerrilla Warfare*. New York: Monthly Review Press, 1961; reprint, Lincoln: University of Nebraska Press, 1985.
Guillen, Abraham. *Philosophy of the Urban Guerrilla: The Revolutionary Writings of Abraham Guillen*. Translated and edited by Donald C. Hodges. New York: William Morrow, 1973.
Guzmán, Abimael. "El Discurso del Dr. Guzmán." In *Los partidos políticos en el Perú*, edited by Rogger Mercado U., 85–90. Lima: Ediciones Latinoamericanas, 1985.
Hakim, Peter. "Is Washington Losing Latin America?" *Foreign Affairs* (January/February 2006): 39–53.
Hammes, T. X. *The Sling and the Stone: On War in the Twenty-first Century*. St. Paul, Minn.: Zenith Press, 2006.
Herring, Hubert. *A History of Latin America*. 3rd ed. New York: Alfred A. Knopf, 1972.
Holloway, John. *Cambiar el mundo sin tomar al poder: El significado de la revolucion hoy*. Buenos Aires: Universidad Autonoma de Puebla, 2002.
Homer-Dixon, Thomas F. *Environment, Scarcity, and Violence*. Princeton, N.J.: Princeton University Press, 1999.

Horne, Alistair. *A Savage War of Peace*. New York: Viking Press, 1978.
International Institute for Strategic Studies. *The Military Balance, 2009*. London: International Institute for Strategic Studies, 2009.
Joes, Antony James. *From the Barrel of a Gun*. Washington, D.C.: Pergamon-Brassey's, 1986.
Jordan, David C. *Drug Politics: Dirty Money and Democracies*. Norman: University of Oklahoma Press, 1999.
Jordan, Javier, and Nicola Horsburg. "Mapping Jihadist Terrorism in Spain." *Studies in Conflict and Terrorism* 28, no. 3 (2005): 174–79.
Judt, Tony. *Post War: A History of Europe since 1945*. New York: Penguin Press, 2005.
Kitson, Frank. *Warfare as a Whole* [1969]. London: Faber and Faber, 1987.
Klepak, Hal. *Cuba's Military, 1990–2005: Revolutionary Soldiers during Counter Revolutionary Times*. New York: Palgrave MacMillan, 2005.
Krasner, Stephen, and Carlos Pascal. "Addressing State Failure." *Foreign Affairs* (July/August 2005): 153–63.
Larteguy, Jean. *The Centurians*. New York: E. P. Dutton, 1961.
Last, David. "Winning the Savage Wars of Peace." In *The Savage Wars of Peace*, edited by John T. Fishel, 211–39. Boulder, Colo.: Westview Press, 1998.
Latell, Brian. *After Fidel: The Inside Story of Castro's Regime and Cuba's Next Leader*. New York: Palgrave MacMillan, 2005.
Legler, Thomas. *Bridging Divides, Breaking Impasses*. FOCAL Policy Paper. Ottawa, Canada: FOCAL, 2006.
Lenin, V. I. *The Lenin Anthology*. Edited by Robert C. Tucker. New York: W. W. Norton, 1975.
Levy, David. "The Vietnam War." *Word, Magazine of the (Cleveland, Okla.) Pioneer Library System* (Spring 2011): 4–5.
Liddell-Hart, B. H. *Strategy*. 2nd ed. New York: Praeger, 1954.
Lupsha, Peter. "The Role of Drugs and Drug Trafficking in the Invisible Wars." In *International Terrorism: Operational Issues*, edited by Richard H. Ward and Harold E. Smith, 177–90. Chicago: Office of International Criminal Justice, University of Illinois at Chicago, 1987.
———. "Towards an Etiology of Drug Trafficking and Insurgent Relations." *International Journal of Comparative and Applied Criminal Justice* (Fall 1989): 60–74.
Mackinley, John, and Alison Al-Baldawy. *Rethinking Counterinsurgency*. Santa Monica, Calif.: RAND, 2008.
Manwaring, Max G. *Gangs, Pseudo-Militaries, and Other Modern Mercenaries: New Dynamics in Uncomfortable Wars*. Norman: University of Oklahoma Press, 2010.
———, ed. *Gray Area Phenomena: Confronting the New World Disorder*. Boulder, Colo.: Westview Press, 1993.

———. *Insurgency, Terrorism, and Crime: Shadows from the Past and Portents for the Future*. Norman: University of Oklahoma Press, 2008.

———. *A "New" Dynamic in the Western Hemisphere Security Environment: The Mexican Zetas and Other Private Armies*. Carlisle Barracks., Penn.: Strategic Studies Institute, 2009.

———. "Security of the Western Hemisphere: International Terrorism and Organized Crime." *Strategic Forum* (April 1998): 2–5.

———. *Street Gangs: The New Urban Insurgency*. Carlisle Barracks, Penn.: Strategic Studies Institute, 2005.

———, ed. *Uncomfortable Wars: Toward a New Paradigm of Low Intensity Conflict*. Boulder, Colo.: Westview Press, 1991.

Manwaring, Max G., and John T. Fishel. "Insurgency and Counter-Insurgency: Toward a New Analytical Approach." *Small Wars and Insurgencies* 3, no. 3 (Winter 1992): 272–310.

———, eds. *Toward Responsibility in the New World Disorder: Challenges and Lessons of Peace Operations*. Portland, Ore.: Frank Cass, 1998.

Manwaring, Max G., and Anthony James Joes, eds. *Beyond Declaring Victory and Coming Home: The Challenge of Peace and Stability Operations*. Westport, Conn.: Praeger, 2000.

Manwaring, Max G., and William J. Olson, eds. *Managing Contemporary Conflict: Pillars of Success*. Boulder, Colo.: Westview Press, 1996.

Manwaring, Max G., and Courtney Prisk, eds. *El Salvador at War: An Oral History*. Washington, D.C.: National Defense University Press, 1988.

Manwaring, Max, Donald E. Shulz, Robert Maguire, Peter Hakim, and Abigail Horn. *The Challenge of Haiti's Future*. Carlisle Barracks, Penn.: Strategic Studies Institute, 1997.

Mao Tse Tung. *On Protracted War*. May 1938. Available online at http://www.marxists.org/reference/archive/mao/selected-works/volume-2/mswv2_09.htm.

———. *On the Protracted War*. Peking: Foreign Languages Press, 1954.

Marighella, Carlos. *The Manual of the Urban Guerrilla*. Chapel Hill, N.C.: Documentary Publications, 1985.

Maritain, Jacques. *Man and the State*. Chicago: University of Chicago Press and Phoenix Books, 1951.

Marks, Thomas A. *Maoist Insurgency since Vietnam*. London: Frank Cass, 1966.

Mathews, Matt M. *We Were Caught Unprepared: The 2006 Hezbollah-Israeli War*. Occasional Paper 26. Fort Leavenworth, Kans.: U.S. Army Combined Arms Center Combat Studies Institute Press, 2008.

McCormick, Gordon H. *The Shining Path and Peruvian Terrorism*. Rand P-7297. Santa Monica, Calif.: RAND, 1987.

Mercado Jarrín, Edgardo. "El impacto de la crisis sobre los conflictos geopolíticos." *Defensa Nacional* (August 1986): 53–83.

———. "Seguridad y fuerzas armadas." *Defensa Nacional* (October 1987): 141–53.

Mercado U., Rogger, ed. *Los partidos políticos en el Perú*. Lima: Ediciones Latinoamericanas, 1985.

Metz, Steven, and Raymond Millen. *Future War/Future Battlespace: The Strategic Role of American Landpower*. Carlisle Barracks, Penn.: Strategic Studies Institute, 2003.

Miller, David C., Jr. "Back to the Future: Structuring Foreign Policy in a Post–Cold War World." in *Managing Contemporary Conflict: Pillars of Success*, edited by Max G. Manwaring and William J. Olson, 13–28. Boulder, Colo.: Westview Press, 1996.

Millett, Richard L. *Searching for Stability: The U.S. Development of Constabulary Forces in Latin America and the Philippines*. Fort Leavenworth, Kans.: Combat Studies Institute Press, 2010.

Millett, Richard, and Thomas Shannon Stiles. "Peace without Security: Central America in the 21st Century." *Whitehead Journal of Diplomacy and International Relations* (Winter–Spring 2008): 31–33.

Nguyen Giap, Vo. *People's War, People's Army: The Viet Cong Insurrection Manual for Underdeveloped Countries*. New York: Frederick A. Praeger, 1962.

Nyrop, Richard F., ed. *Guatemala: A Country Study*. Washington, D.C.: American University and Department of the Army, 1983.

Obando, Enrique. "Defeating Shining Path: Strategic Lessons for the Future." In *Saving Democracies*, edited by Anthony James Joes, 125–43. Westport, Conn.: Praeger, 1999.

Olson, William J. "International Organized Crime: The Silent Threat to Sovereignty." *Fletcher Forum* (Summer/Fall 1997): 75–78.

Palmer, David Scott. "The Sendero Rebellion in Rural Peru." In *Latin American Insurgencies*, edited by Georges Fauriol, 67–96. Washington, D.C.: National Defense University Press, 1985.

Passage, David. "Reflections on Psychological Operations: The Imperative of Engaging a Conflicted Population." In *Ideas as Weapons*, edited by G. J. David, Jr., and T. R. McKeldin III, 49–58. Washington, D.C.: Potomac Books, 2009.

Payne, Douglas W. "The Drug Super State in Latin America." *Freedom at Risk* (March–April 1989): 7–10.

Petraeus, David H., James A. Amos, and John A. Nagl. *The U.S. Army/Marine Corps Counterinsurgency Field Manual*. U.S. Army Field Manual no. 3-24 and Marine Corps Warfighting Publication no. 3-33.5. Chicago: University of Chicago Press, 2007.

Pizarro, Eduardo, and Ana Maria Bejarano. "Colombia: A Failing State?" *ReVista: Harvard Review of Latin America* (Spring 2003): 1–6.

Qiao Liang and Wang Xiangsui. *Unrestricted Warfare*. Beijing: PLA Literature and Arts Publishing House, 1999.

Rapley, John. "The New Middle Ages." *Foreign Affairs* (May/June 2006): 93–103.

Research Institute for the Study of Conflict and Terrorism. *Shining Path: A Case Study in Ideological Terrorism*. Case Studies in Terrorism no. 260. London: Research Institute for the Study of Conflict and Terrorism, 1993.

Ricks, Thomas E. *Fiasco: The American Military Adventure in Iraq*. New York: Penguin Press, 2006.

Robinson, Linda. *Tell Me How This Ends: General David Petraeus and the Search for a Way Out of Iraq*. New York: Public Affairs of Perseus Book Group, 2008.

Ropp, Steve. *The Strategic Implications of the Rise of Populism in Europe and South America*. Carlisle Barracks, Penn.: Strategic Studies Institution, 2005.

Sanchez, W. Alejandro. "The Rebirth of Insurgency in Peru." *Small Wars and Insurgencies* (Autumn 2003): 185–98.

Schwirtz, Michael. "Russia's Political Youths." *Demokratizatsya* (Winter 2007): 80–81.

Sendero. "El documento oficial del Sendero." In *Interview with Chairman Gonzalo*, edited by Abimael Guzmán Reynoso. San Francisco, Calif.: Red Banner Editorial House, 1988.

Smith, Paul E. *On Political War*. Washington, D.C.: National Defense University Press, 1989.

Smith, Rupert. *The Utility of Force: The Art of War in the Modern World*. New York: Alfred A. Knopf, 2007.

Steinitz, Mark S. "Insurgents, Terrorists, and the Drug Trade." *Washington Quarterly* (Fall 1985): 147.

Strong, Simon. *Shining Path: Terror and Revolution in Peru*. New York: Random House, 1992.

Sullivan, John P. "Maras Morphing: Revisiting Third-Generation Gangs." *Global Crime* (August–November 2006): 493–94.

———. "Terrorism, Crime, and Private Armies." *Low Intensity Conflict and Law Enforcement* (Winter 2002): 239–53.

Summers, Harry G., Jr. *On Strategy: A Critical Analysis of the Vietnam War*. New York: Random House, 1982.

Sun Tzu. *The Art of War*. Translated by Samuel B. Griffith. 1963; reprint, Oxford: Oxford University Press, 1971.

Tarazana-Sevilleno, Vanda. *Sendero Luminoso and the Threat of Narcoterrorism*. New York: Praeger, 1990.

Taw, Jennifer M., and Bruce Hoffman. *The Urbanization of Insurgency: The Potential Challenge to U.S. Army Operations*. Santa Monica, Calif.: RAND, 1994.

Toffler, Alvin, and Heidi Toffler. *War and Anti-war: Survival at the Dawn of the Twenty-first Century*. New York: Little, Brown, 1993.

Torres, Vladimir. *The Impact of "Populism" on Social, Political, and Economic Development in the Hemisphere.* FOCAL Policy Paper. Ottawa, Canada: FOCAL, 2006.

Trinquier, Roger. *Modern Warfare: A French View of Counterinsurgency.* Barcelona, Spain: n.p., 1965. English translation, Fort Leavenworth, Kans.: Combat Studies Institute, 1964.

Ungo, Guillermo M. "The People's Struggle." *Foreign Policy* (Fall 1983): 51–63.

U.S. Army Peacekeeping and Stability Operations Institute. *Guiding Principles for Stabilization and Reconstruction.* Carlisle, Penn.: U.S. Army Peacekeeping and Stability Operations Institute, 2009.

Verstrynge Rojas, Jorge. *La guerra periferica y el Islam revolucionario: Origines, reglas, y etica de la guerra asimetrica.* Madrid: El Viejo Topo, 2005.

Vidino, Lorenzo. *Al Qaeda in Europe.* Amherst, N.Y.: Prometheus Books, 2006.

Villalobos, Joaquin. "El estado actual de la guerra y sus perspectivas." *ECA Estudios Centroamericanos* (March 1986): 169–204.

Williams, Phil. *From the New Middle Ages to a New Dark Age: The Decline of the State and U.S. Strategy.* Carlisle Barracks, Penn.: Strategic Studies Institute, 2008.

World Bank. "Crime, Violence and Development: Trends, Costs, and Policy Options in the Caribbean." Report No. 37820, Joint Report by UN Office on Drugs and Crime in the Latin American Region of the World Bank, March 2007.

Yin, Robert K. *Case Study Research: Design and Methods.* 2nd ed. Thousand Oaks, Calif.: SAGE Publications, 1994.

Index

Afghanistan, 64, 68, 125, 142, 160, 161, 162, 163, 164, 165, 166
Africa, 48, 57, 61, 62, 109
Agi-prop efforts. *See* Propaganda war
Ait Ahmed, Hocine, 13, 15–16
Algeria, x, 10–11, 54; revolution in, 12–16, 18, 124
Al Qaeda: asymmetric warfare and, ix, x, 51, 52, 56, 140, 149, 161–62; organization of, xiv, 54, 57; in Spain, xi, 53, 54–57, 148
Andean Ridge, 42, 49, 109
Arab states, 15, 147
Argentina, ix, 31
Aristide, Jean-Bertrand, 64, 65, 66
Arms trafficking, 71, 100, 104, 109, 110, 112
Asia, 15, 48, 88, 109, 162, 167
Assassination, 27, 35, 36, 42, 81, 98, 105–106, 111, 162
Asymmetric war, ix, xiv, 3, 6, 16–18, 134–35, 148–51, 161–67; Al Qaeda and, 51, 52, 56, 140, 149, 161–62; case study methodology and, 5–6, 155–57; nonstate actors and, 49, 55, 67, 73–74, 78, 108–109, 123–25, 133, 135, 143, 149–50; objectives of, 43, 77, 90, 120–22, 133, 140; responses to, ix, 13–14, 18, 22, 23, 46–48, 61, 117, 129–32, 149, 155–66; success in, 13, 18, 45–46, 126, 145–46, 157–58, 163, 165–67; types of, 80, 88, 89, 92, 120, 122, 126–28, 141–45

Belize, 104, 115
Biological war, xiii, 6, 120, 122, 126, 127, 129, 133–34, 144; defending against, 129–32
Bolívar, Simón, 31, 41
Bolivia, viii, 92, 160
Brazil, xi, 53, 69–73, 141, 180n40

Cacos, 53, 63, 64, 65, 66, 69
Castro, Fidel, 53, 98, 179n23
Center of gravity, 4, 9–10, 11, 15, 24, 26–27, 46, 99, 138; asymmetrical war and, 129, 138–39, 145–46, 149–50; demonstrations and, 80, 81, 82, 84, 85, 86–88, 91, 95; human resources and, 52, 66, 74, 116, 118, 124, 134, 161; human terrain and, 16–17, 43, 52, 139, 150; propaganda war and, 13, 36, 45, 49–50, 55, 56, 59, 68, 139–40, 146–47; public opinion as, 77, 95, 114; terrorism and, 74
Central America, vii–ix, 28, 103, 104, 115–16, 141, 161
Chavez, Hugo, ix, 4, 5, 137, 170n6, 178n2, 190n3
China, xiii, 15, 124, 134, 161, 167
Clausewitz, Carl von, xiv, 9, 13, 14, 16–17, 30, 160, 161, 190n5

Cold war. *See under* War
Colom, Alvaro, 102, 103, 104, 105–106, 107
Colombia, viii–ix, xi, xii, 41, 42, 142, 160, 165, 166
Communism, 15, 44, 48, 78–79, 81–82, 98, 122, 123–24, 162
Counterinsurgency, vii, xi, 13, 14, 18, 22, 46, 48, 117, 126, 155–61
Cuba, xi, xii, 22, 23, 25, 28, 98, 106, 124; popular militias (paramilitary forces) in, 52, 53, 57–61, 77
Cyber war: contemporary battlefield and, xiii–xiv, 6, 17, 75, 80, 85, 95, 120, 121, 122, 126–30, 133–34, 140, 144, 167; defending against, 129–32; Russia and, 88–93, 127–28

Democracy: direct, 46, 114; electoral process and, 110, 111, 112, 113, 116; market capitalism and, 45, 89; reform and, ix, 78, 101, 124; "true," 33, 40, 47, 48, 78–79. *See also* Social democracy
Diplomatic war: in Algeria, 13, 14, 15, 18; contemporary battlefields and, 11, 17, 18, 48, 80, 95, 131, 139–40, 144, 147, 150; in Latin America, 21, 26, 27
Divine surprise, 5, 53, 89, 148–49; asymmetric war and, 121, 122, 144, 145; in Brazil, 69–70, 71, 72, 73; Russia and, 77, 88–89, 90, 94
Drugs: in Brazil, 71–73; in Central America, vii, 104; in Colombia, vii–ix, xi, 42, 100, 118, 166; gangs and, vii, 65, 66–67, 71, 73; in Guatemala, 100, 103, 104, 105, 112, 118; in Haiti, 65–67; insurgency and, vii–ix, 42, 49, 115; in Mexico, vii, 104, 115, 118; organized crime and, vii, 67, 73, 109; in Peru, xi, 36–37, 38, 41–42; political change and, 6, 71, 72, 98, 111–12, 141, 143; in the United States, vii, 42, 104, 115, 132. *See also* Narcotrafficking
Duarte, José Napoleón, xi, 28, 48, 165

Economic war, 17, 45, 68, 69–70, 75, 79–81, 89, 90–91, 97, 131, 139, 141, 144

El Salvador, 92, 155–56, 159, 160–61, 164–65; armed forces of, x–xi, 19–21, 24–25, 26, 47, 48, 172n29; democratic reforms and, xi, 20, 24, 26–28, 47–48; insurgency in, x, xii, 10, 12–13, 19–29, 161, 172n29, 173n53
End-state planning, 11, 18, 142, 152
Estonia, xiii, 83, 87, 88–89, 90–92, 127, 133
Europe, 12, 14, 86, 158, 167; Al Qaeda and, x, 52, 53, 54–57, 148, 162; communism/socialism and, 15, 32, 123, 124; drug trafficking and, viii, 42; Eastern, 15, 124; political instability in, 14, 51, 148; terrorism in, 48, 53, 54, 148, 162; Western, 15, 52, 54, 56, 124, 148

Farabundo Martí National Liberation Front (FMLN), 11–12, 20, 22–23, 25–29
Focos, 11, 27, 31
France, x, 12–15, 26, 62–63, 64, 65, 162
Frente Amplio, 113, 114, 115
Fuerzas Armadas Revolucionarias Colombianas (FARC), viii, ix, xi, 41, 42, 166

Gangs: asymmetric war and, 61, 71, 79–81, 143; in Brazil, 70–73; drug trafficking and, vii–viii, xii, 65, 66–67, 71, 73, 105, 110, 115–16, 141; in Guatemala, 100–101, 102, 104, 107, 110, 111–12, 115–17; in Haiti, 52, 53, 65–67, 69; insurgency and, 117, 137; nonstate actors and, x, 4, 51, 52, 57, 112; state actors and, ix, xi–xiii, 4, 6, 65–66; youth leagues as, 76–77, 81–82, 94–95, 114
Georgia, xiii, 83, 88–89, 90, 92–93, 127, 133
Germany (East, West, and unified), 64, 87, 123, 188n12
Giap, Vo Nguyen, 25, 26, 40
Guatemala: civil war in, 98–99, 118; ex-*combatantes* in, 99, 101, 103; Frente Amplio of, 113, 115; gangs in, xii, 97, 104, 107, 115–16, 118; insurgency in, viii, 97, 99–100, 113, 185n8; legitimacy of, 105–107, 109–10,

116–17; politics in, 98–100, 102–107, 110–11, 113; revolution in, 112–16, 118; sovereignty of, 100–104, 107–10, 118–19; URNG in, 99, 185n8
Guerrilla war, 49–50, 137, 148; in Afghanistan, 166; in Algeria, 13; in Brazil, 71; in Colombia, ix, 166; in Cuba, 58, 60; in El Salvador, 11, 22–25, 27, 28–29; in Guatemala, 98–99, 185n8; as "kindler, gentler" war, 6, 44, 49, 113, 191n15; in Peru, 34, 39–44, 160; revalidation of, 4, 30–31, 38–39, 44, 49, 137, 144; in Vietnam, 161, 162
Guevara, Ernesto "Che," 11, 20, 25, 26, 28–29, 31, 35
Guillen, Abraham, 39, 43, 113, 141, 142
Guzmán, Abimael, xi, 37, 40, 45, 49; five-stage plan of, 32, 33–37; sixth stage of, 38–44

Haiti: earthquake in, 53, 66; gangs in, 52, 65, 66–68, 69, 141; history of, 61–65; United States and, 61, 64–65, 66
Hezbollah, 145–48
Holloway, John, 98, 112, 113, 116
Huallaga Valley, xi, 36, 41
Human rights, 27, 30, 64, 70, 99, 110, 111
Human terrain, 16–17, 43, 52, 139, 150
Human trafficking, 71, 100, 104, 109, 112, 157

Impunity, 65, 67, 72, 73, 74, 99, 101, 106, 107, 112, 118
Information war: in Algeria, 13, 14, 15, 18; contemporary battlefields and, 16, 17, 18, 29, 48, 52, 59–60, 80–81, 95–96, 105–106, 144; counterterrorism and, 126, 130, 132; cyber-attacks as, 85, 91–92, 120, 122, 126, 128–29, 131–32, 133; in El Salvador, 27; by Nashi, 85, 90–92, 95; in Peru, 41, 45–46; soft power and, 48, 114, 129, 139–40. *See also* Propaganda war
Insurgency, 76, 137, 138–41; in Algeria, x; in Colombia, viii, 160; countering, 45–46, 116–17, 156–67; in El Salvador, viii, x–xi, 11–13, 19–29; financing, 55; in Guatemala, 98–99, 103–104, 116–19; in Haiti, 65–67; in Iraq, 161–63; narcotrafficking and, vii–ix, 42, 49, 66–67, 103; in Peru, 33, 41; state failure and, 109, 116–17, 119; super, 5, 137; in Uruguay, ix; victory and, x, 13–14, 18, 67, 161
Intelligence war, 11, 80, 95, 130, 139–40, 145, 146–47, 152
Internet, 54, 90, 92, 111
Iran, xiii, 162, 167
Iraq War, 56–57, 64, 68, 125, 142, 160, 161–63, 164, 165
Islamic militants, 5, 12, 51, 52, 55, 56, 122, 140, 147, 150, 162
Israel, xiii, 145–48

Kasparov, Gary, 87–88
Kidnapping, 27, 67, 71, 80, 81, 100, 109, 111
Kinetic (military) power: contemporary battlefield and, 18–19, 60, 77, 134, 143, 146–47; vs. nonkinetic power, 11, 80, 90, 93, 95, 139; as traditional paradigm, 14, 17, 22–23, 43, 68, 144, 145–47
Komsomol, 81–82, 188n12

Latin America: drug trafficking and, 104, 109, 115–16; OAS and, 65; political systems in, 21, 30, 31, 108–109; revolution in, 11, 44, 49, 148, 157, 161
Law enforcement: biological/cyber war and, 131; gangs and, 74; paramilitaries and, 85; political stability and, 72–73, 100–101, 105–107, 108, 117–18, 139; terrorism and, 55, 68, 74, 80, 118
Legitimacy: challenges to, 11, 16, 19, 35, 69, 105–107; insurgency and, 24–25, 27, 28, 29, 40–41, 46, 103, 165–66; moral, 40, 57, 116–17, 118–19; psychological war and, 20, 31, 33, 43, 68, 150, 151; public services and, xii, 27, 49, 59, 101
Lenin, Vladimir Ilyich, ix, 10–11, 29, 30–31, 32, 40, 41, 45, 81, 142; as dictator, 59, 76

Leninist revolution, viii, ix, 4, 10, 20, 28–29, 30, 35, 38, 51, 59, 60, 137; democracy and, 33, 40, 77, 78–79; militias/gangs and, 74, 78; process of, 79–81; vanguard of the proletariat and, 34, 40–41
Liddell-Hart, B. H., 10, 13, 96, 142

Manwaring Paradigm, 156–58, 161, 165–66, 167
Maoist revolution, viii, 35
Mao Tse Tung, 4, 25, 31, 38, 40
Maras, vii, 100, 103, 104, 110, 115, 118
Mara Salvatrucha (MS-13), xiii, 100, 141
Marighella, Carlos, 125
Marxist revolution, viii, ix, 20, 30, 33, 34, 78, 123, 124, 142
Maya Biosphere Reserve, 103–104
Media war, 4, 11, 17, 59, 68, 75, 80, 139, 144, 147, 151
Mexico, 26, 167; drug trafficking and, vii, ix, 100, 104, 106, 115, 118; gangs of, vii, viii, xii, 100, 103, 104, 110, 115, 141
Middle East, 48, 55, 109, 148
Militias, popular, xii, 4, 60, 61, 74; in Brazil, 73; in Cuba, xi, xii, 52, 57–61. *See also* Paramilitary organizations
Military force. *See* Kinetic war
Multidimensional paradigm: asymmetric war and, 4, 10, 13, 17, 27, 31, 38, 138–39, 141–42; for countering asymmetric war, 13, 147, 150–51, 152
Murder, 64, 67, 71, 73, 80, 84, 87, 100, 105–106, 109

Narco-states, 46, 104, 110–12, 116, 141, 160
Narcotrafficking, 141, 158, 166; arms trade and, 54, 71, 98, 100, 104, 112; corruption and, 65, 100, 106, 110–12; gangs and, vii–ix, 105, 108–109, 112, 115–16; in Haiti, 65, 66; in Latin America, vii–ix, 37, 71–72, 98, 104, 160, 166; narco-insurgent-paramilitary alliance and, 41–42, 66, 100–101, 105, 108–109

Nashi, 81–82, 90, 92; paramilitary training of, 84–85, 95; political activities of, 84, 85–88, 93–94, 95; as youth gang, xii, 59, 61, 76–78, 82–83, 94
Nation-states: allies and, 14–15, 47; challenges to, xii–xiii, 11, 20, 35–36, 39, 53, 71, 78, 82, 136–40, 148, 149–50, 158; civil-military juntas and, 20, 21, 61, 160, 172n29; coercion and, 6, 11, 31, 50, 60, 77, 80–81, 109, 111, 113, 141; contemporary battlefields and, 7–8, 13, 60, 68–69, 140–41; corruption and, 33, 66–67, 72, 80, 100–101, 102, 105, 107, 108, 111, 117; criminal free-states in, 42–43, 52, 66–67, 72, 73, 100, 103, 104, 109–12, 115–16; internal war and, 51, 76, 98, 121–22, 137–38, 143; legitimacy and, 11, 14, 19, 23, 24–25, 27, 33, 49, 57, 101–102, 103, 105–109, 114–19, 146, 150, 165–66; militias and, 52, 57, 58–60, 64; nationalism and, 84, 93–94; political instability and, 63, 65, 89–90, 97, 111–12; political process and, 31, 64–65, 83–84, 114–15; poverty, 61, 65, 72, 109; public services and, xii, 33, 59, 64, 67, 69–70, 89, 108–109; rivalry and, 15–16; state failure and, 43, 46, 47, 50, 67, 69, 102, 107–10, 118–19, 125; state-supported gangs and, 76, 77, 80–84; war and, 128–29, 136, 143
Neopopulists, 4, 93, 137
New Left, 4, 98, 113, 114, 116, 123–25, 133, 150, 188n12
New Socialists, 4, 44, 93, 116, 124–25, 133, 137, 142, 150
Nicaragua, 20, 22, 23, 25, 28, 161
9/11. *See* September 11, 2001
Nongovernmental organizations (NGOs), 76, 98, 99, 100, 141, 158
Nonstate actors: asymmetric war and, 3, 6–7, 67, 75, 76, 122, 126, 129, 133, 138–43, 145–49, 152, 153, 167; coercion and, 11, 68, 70, 77, 79, 90, 139, 141; criminal, 4, 66–67, 69–70, 72–73, 116, 118, 119; insurgency and, 13, 41–42, 49, 52, 57, 59; state failure and, 78, 82, 109, 110, 111–12, 149
North Africa, 14–15, 54, 55, 148

North Atlantic Treaty Organization (NATO), 15, 90, 93, 163
North Korea, 167
North Vietnam, 23, 25, 40

Old Left, 123, 124, 125–26, 133
Organized crime. *See* Transnational criminal organizations

Pandillas, 100, 118
Paramilitary organizations, viii–ix, 53, 57–61, 84–85. *See also* Militias, popular
People's Liberation Army (PLA), 39, 43, 44, 45–46, 49
People's Revolutionary Army (ERP), 22, 23
People's war, 5, 34, 40, 48, 58, 68, 77, 113, 136–39, 141
Peru, viii, 30, 140, 160; armed propaganda in, 37–38, 43–44, 45, 109; continental revolution and, 41–42, 45; democratic reforms in, 32, 43, 45–47; drug trafficking and, 36–37, 42–43; Guzmán's action plan for, 33–37, 39–41; history of, 31–33, 35. *See also* Sendero Luminoso
Popular Front, 39, 41, 60, 113
Populists, 4, 108, 137
Poverty, 61, 65, 72, 94, 109
Prada, Manuel González, 31–32
Primeiro Comando da Capital (PCC), 53, 69–73
Propaganda war, 18, 59, 68, 79–81, 95, 139, 146, 150–51; in Algeria, 13, 16; Hezbollah and, 145–47; Nashi and, 79, 81–82, 84–86, 95; as soft power, 31, 43, 45, 50, 52, 55, 76, 139–40; violence and, 36, 38, 56, 74. *See also* Center of gravity; Information war
Putin, Vladimir, 82–83, 84, 85, 87, 93, 94

Qiao Liang, 13, 17, 50, 74, 88, 89, 120, 138, 142, 144, 191n15

Red Army Faction (RAF), 123
Red Brigades, ix, 48, 123–24
Revolution, 123–25, 138–39, 141, 148; in Algeria, 12–19; in Cuba, 57–59; in El Salvador, 22–29; in Guatemala, 108, 112–16; in Latin America, 41–42, 49, 97–98; legitimate governance and, 59, 82, 97–98, 114; Leninist, ix, 11, 34–35, 43, 78, 79–81, 114; military, ix, 19, 20, 27, 28; paramilitaries and, 59–60; in Peru, 33–37, 39–44; in Russia, 59; in Vietnam, 40; worldwide, 42, 44–45, 78–79, 82, 149
Rosenberg Marzano, Rodrigo, 105–106
Russia (former Soviet Union), 162, 167, 188n12; cyber-attacks and, xiii, 88–93, 133; defense of, xii, 77, 81, 82–84, 85–86, 93, 94, 134; politics in, 82, 83–84, 85–88, 94; youth leagues of, xii, 59, 61, 76, 81–83, 84–85, 90, 93, 94

Sendero Luminoso (Shining Path), xi, 30, 37–38, 45, 47, 49, 140; drug trade and, 36–37, 38, 41–42; organization of, 34; strategic plan of, 32–37, 39, 43, 48; violence of, 38, 40, 109, 160
Sepoyan militarism, 40, 59
September 11, 2001 (9/11), xiii, 89, 121, 161
Shining Path. *See* Sendero Luminoso
Social democracy, 33, 44, 46, 59, 60, 77, 78
Socialism, 78, 82, 89, 94, 142; Cuban, 57, 59, 60; European, 32, 56; Latin American, 40, 44, 108, 112, 114, 141
Soft (nonkinetic) power, 18, 31, 45–48, 49, 60, 80, 90, 94–95, 139
Southeast Asia, 12, 15, 29, 48, 160, 161, 162
South Ossetia, xiii, 89, 92–93
Soviet Union, 15, 23, 25, 28. *See also* Russia
Spain, xi, 4, 53, 54–57, 148, 149
Sun Tzu, 4, 10, 29, 30, 74, 139, 165
SWORD model, 155–56, 158, 159

Territory, control of, xii, 4, 16, 23, 26, 34, 35, 52, 66, 133; gangs/TCOs and, 71, 73, 101, 103, 109; war and, 69, 96, 138
Terrorism: acts of, 27, 54–56, 80–81, 126–27, 148, 161; goal of, 74, 133, 137; nonstate actors and, 41, 44, 52, 53, 65, 74, 133, 148; state failure and, 109, 111–12; state response to, 15,

Terrorism *(continued)*
46, 55, 56, 57, 64, 131; WMDs and, 121, 126
Trade unions, 40, 80, 113, 114
Transnational criminal organizations (TCOs): arms trafficking and, 71, 100, 104, 109, 110, 112; asymmetric war and, 141, 143; gangs and, vii–ix, xii, 66, 71–73, 115–16; in Guatemala, 97, 100–101, 103, 107, 111–12, 115–16, 118; human trafficking and, 71, 100, 104, 109, 112, 157; insurgency and, 41, 117; terrorism and, 54, 137
Trojan horses, 6; asymmetric war and, 5, 53, 77, 94, 122, 144, 148–50; zones of impunity and, xi–xii, 51–52, 67, 73, 74, 93
Tupamaros, ix, 115

Ukraine, 82, 85, 93
Uncomfortable wars phenomena. *See under* War
United Nations (UN), 15, 47, 65, 98, 99, 100, 106–107, 117, 152
United States, 12, 15, 34, 87; asymmetric warfare and, ix, x, 3, 22, 48, 117, 125–28, 137, 148, 162, 162, 167; Cuba and, 58; drug trafficking and, viii, 3, 42, 104, 117, 132, 160, 166; economic aid and, 21, 24, 26, 47, 164, 166; El Salvador and, 19, 21, 23, 24, 25, 26, 27, 47, 164–66; Haiti and, 61, 64; as imperialist, 40, 56–57, 59–60, 114; Latin America and, vii–viii, xii–xiii, 40, 47, 157, 161, 166; Middle East and, 160, 161–64, 166; military of, 64, 65, 158, 162–63, 164; Russia and, 23, 85, 87, 92, 134; security and, xii, xiii, 22, 55, 56, 57, 64, 117, 126, 127–28, 132, 133, 134, 142, 152, 160–62, 164–67; Vietnam War and, 23, 25, 26, 160, 161, 162, 164, 165
Unity of effort, 17, 18, 21–23, 25, 29, 146, 147, 151
Uruguay, ix, xi, 28, 115

Vanguard of the proletariat, 20, 34, 39–41, 45, 114
Venezuela, ix, 5, 31, 41–42, 49, 61, 137, 170n6, 178n2, 190n3. *See also* Chavez, Hugo
Verstrynge Rojas, Jorge: asymmetric/guerrilla war and, 4, 10, 30–31, 38, 39, 43, 44, 120, 121, 122, 134, 137, 144, 170n6, 190n3; divine surprises and, 88, 89, 122, 148, 150; social disequilibrium and, 51, 148
Vietnam War, x, 11, 12, 23, 29, 40, 48, 124, 151, 160, 161, 162, 164, 165

Wang Xiangsui, 13, 17, 50, 74, 88, 89, 120, 138, 142, 144, 191n15
War, 191n15; case study methodology and, 5–6, 155–57; civil, 98–99, 100, 160; cold, 3, 6, 9, 10, 161; contemporary, 45, 50, 68–69, 74–75, 76, 95–96, 136–40, 143, 151–52; dimensions of, 4, 10, 11, 17, 27, 29, 31, 95, 138–39, 144–45, 147, 151, 152, 161; fourth-generation war (4GW), 4, 5, 170n6, 178n2, 190n3; hegemonic, 3, 9, 142, 143, 150, 153, 158, 162; insurgency as, 30–31, 38–39, 49–50, 67, 98, 161–62; interstate, 7, 143, 167; intrastate, 51, 76, 98, 121–22, 137–38, 143, 152; irregular, 16, 55, 122, 133, 142, 145, 166–67; nonstate actors and, 3, 6–7, 49, 51, 67, 75–78, 122, 126, 129, 133, 138–40; objective of, 38–39, 43–44, 67, 90, 95, 103, 143; political change and, 7, 17, 23, 60, 77–78, 95, 103, 122, 133, 137, 140–41; protracted, 4, 20, 34, 38, 39, 43–44, 80; social instability and, 16–17, 27, 49–50, 52, 77, 78, 89, 95; strategic questions and, 9–10; total, 10, 13, 17–19, 21, 25, 43, 45, 69, 89–90, 121–22, 133, 134, 141; uncomfortable, ix, xi, 138, 155–59, 160, 167; victory and, 17, 21–22, 96, 120, 122, 142, 156, 157–58, 163, 167. *See also* Asymmetric war
Weapons of mass destruction (WMDs), 109, 120–21, 126, 134, 162
World War II, 12, 14, 32, 86, 87, 90, 128, 156, 163

Zetas, 100, 103, 104, 110, 115, 141

www.ingramcontent.com/pod-product-compliance
Lightning Source LLC
Chambersburg PA
CBHW031643170426
43195CB00035B/402